Laboratory Scientific
Glassblowing
A Practical Training Method

Laboratory Scientific
Glassblowing
A Practical Training Method

Paul Le Pinnet

SOG Ltd, UK

World Scientific

NEW JERSEY · LONDON · SINGAPORE · BEIJING · SHANGHAI · HONG KONG · TAIPEI · CHENNAI · TOKYO

Published by

World Scientific Publishing Europe Ltd.
57 Shelton Street, Covent Garden, London WC2H 9HE
Head office: 5 Toh Tuck Link, Singapore 596224
USA office: 27 Warren Street, Suite 401-402, Hackensack, NJ 07601

Library of Congress Cataloging-in-Publication Data
Names: Le Pinnet, Paul.
Title: Laboratory scientific glassblowing : a practical training method /
  by Paul Le Pinnet (SOG Ltd, UK).
Description: New Jersey : World Scientific, 2017.
Identifiers: LCCN 2016038439| ISBN 9781786341976 (hc : alk. paper) |
  ISBN 9781786342423 (pbk : alk. paper)
Subjects: LCSH: Glass blowing and working. | Glass manufacture. | Glass craft.
Classification: LCC QD63.G5 L4 2017 | DDC 745.2028/2--dc23
LC record available at https://lccn.loc.gov/2016038439

British Library Cataloguing-in-Publication Data
A catalogue record for this book is available from the British Library.

Copyright © 2017 by World Scientific Publishing Europe Ltd.

All rights reserved. This book, or parts thereof, may not be reproduced in any form or by any means, electronic or mechanical, including photocopying, recording or any information storage and retrieval system now known or to be invented, without written permission from the Publisher.

For photocopying of material in this volume, please pay a copying fee through the Copyright Clearance Center, Inc., 222 Rosewood Drive, Danvers, MA 01923, USA. In this case permission to photocopy is not required from the publisher.

Desk Editors: V. Vishnu Mohan/Mary Simpson

Typeset by Stallion Press
Email: enquiries@stallionpress.com

Printed in Singapore

Acknowledgements

I wish to record my heartfelt thanks and appreciation to the following people for their help and support while writing this training manual. Mr John Lewis, Managing Director of SOG Ltd. for his support in this venture and for providing the facilities with which to do so. Eileen Miller for her guidance and publishing knowledge. Charles Davies, Director of SOG Ltd., Konstantin Kraft-Poggensee, (Master) past Chairman, German Scientific Glassblowing Society, Germany, and Mr S. Potter for their support from day one. David B. Hough for his photographic skills. Colin Martin for keeping my computer working and for showing me how to switch it on. Eileen Erlis, Lauren Miller and Julie Carmichael for their typing skills and for making sense of my handwritten script. Alyn Bear for his computer design skills. Leslie James Crampton for offering to buy a copy of the book no matter what the cost. Terry Bratt for providing practical information regarding electrical installation standards. Sandra Cafferty and Louise Dinnage, AKZO Nobel UK Ltd. Dr Carl Waterson MRSC regarding Schlenk Lines. Gary Lloyd, process engineer regarding column packing. Philip Jeffs of Warrington Museum. I would like to record my appreciation for the supportive comments on hearing of the project from my colleagues, friends and customers. Michael Baumbach, Gary Coyn, William Fludgate, Terri Adams, Ian Pearson, Keith Holden, Paul Rathmill, Phil Murray, Jeremy Bolton, Christine Fuhr SCHOTT AG.

Finally "Behind every successful man there is................a surprised woman". My wife Michelle for proof reading my first scripts and for assuring me that the book is understandable to both Technical and Non-Technical readers.

Paul Le Pinnet
24 May 2016

Foreword

I have known Paul Le Pinnet for more years than I can even think about and I have always, in my early years, thought of him as a Mentor and always as a friend, whether I could be considered a Mentor might take me a few more hours at my work bench in the medium that we both work with.

I am still considered to be the apprentice in the eyes of Paul and many of our friends as I came into our profession as a boy.

Let me explain, I have only been Scientific Glassblowing for the last forty two years, I am sure he beats me by a few years, and that does not mean that I know everything there is to know about our skill set, but, a book like this does give me and many like me the information required to complete a particular project, or, another way at solving or at least, how one can look at providing a way to get the end result.

We met many years ago at a meeting of The British Society of Scientific Glassblowers (BSSG), it was whilst talking with Paul and some of "his" friends that I was drawn deeper into our wonderful world of glass. The ability to fight the forces of melting glass and gravity, mixed with centrifugal forces and applying pressure to stop deformation before the glass cools down too soon, or overheating and the flame pushing against the hotspot will destroy hours of work. And the worst enemy of all, the sound that we all try to avoid, the crack!

How to make certain items that would still fox someone outside our magical clique, a glass ship in a bottle, a single surface vessel that is known as a Klein bottle. These were just things to make when the time allows, or as my Mentors would say, practice pieces that need hand, eye and muscle memory to complete, plus the ability to look at a drawing and break the plans down so that we can work from the inside out. Sounds crazy, working from the inside out, but I am sure that the book and its contents will give the insight to what that means.

While the book can almost show the impossible in glass, Paul always insists that it is practice that is the key.

Technical drawings that have been drawn up by engineers and "converting" that into a working breakdown method for us to solve can sometimes be the glassblower's nemesis. Working with different temperatures and pressures and the understanding of what can be pushed upon us by those that require "their" one off specials, gives the glassblower some freedom due to the very nature of the strengths and weakness of that substance they call glass. This comes with the experience and he has kindly drafted some of those mystical practices in these pages.

A few sentimental words about some of my Mentors; they made me burn my hands, cut my fingers, scorch my lips, pick up hot glass and taught me the laws of physics and how glass can tell lies. Let me explain, I would pick up a piece of glass that was very hot and then instantly drop it. It was explained to me that the same piece of glass that was cold must have been lighter and when it was very hot it was heavier due to the natural order of physics. As energy can neither be created nor destroyed, by applying heat to glass, it must have absorbed the energy, exciting the molecules and showing that as heat. As the glass has no feelings, but we do, we can sense that the excited glass must be heavier due to absorption, therefore we drop the heavy glass. Glass on the bench tells you a lie that it is not hot and begs you to pick it up because the "heat" is invisible. That was the way I was taught not to pick up hot glass, because I did not pay attention to my Mentors.

With hundreds of little tips that Paul shares with you, you might just find that you have got away with the pain that my early Mentors thrust upon me.

I would like to say a big thank you to Alex Kovaks and Cecil W. Cullingford for all the laughs I provided you with in my formative years.

I was told by many glassblowers, including the author, "do the best you can, then do it again but make it better, go back and think how it can be made simpler".

I am sure that this book and the way the text is laid out to you, shows that it will pay off, if you put the hours in with practice.

This book is long overdue, and I feel that the author has the wealth of experience and no doubt the hours of work at the bench and behind the hot lathe in the summer months have given a very enjoyable way of life. I must say that when I told my parents that I saw a glassblower at the London Science Museum on a school trip, I was mesmerised by the skill and how easy that James Frost made it look. I contacted him at Reading University when I passed my first glassblowing exam in 1974 and wrote to thank him for his words of wisdom. "Pull spear points until you are bored, pull more spear points until it's no longer boring". He was still working until he was 92 and remained in touch when he finally retired at 95.

We have met with politicians, eminent scientists and of course new people coming into our profession. His enthusiasm shines through and is a credit to all that know him.

I have worked with Paul on committees within the BSSG, be it Council or Board of Examiners, and he always brings to the table common sense, humour and a valuable wealth of experience.

I have attended many of the lectures that he has given and I can always say that I learn something new, usually obscure, but always given in a humorous fashion.

One of the proudest days in my profession was to award and bestow the title of Fellow of the British Society of Scientific Glassblowers to the wizard of words on glass Paul Le Pinnet FBSSG.

One of the things that I have always found with the author, there is no easy way to the top of the ladder in our profession, but I will always be at the bottom holding his ladder, as he has given many people that leg up onto the first rung. I hope that this book brings to you the reader, how we can have the pleasure of understanding the fundamental aspects of our mystical trade.

William Fludgate FBSSG
BSSG Chairman

Introduction

The aim of this manual is to help and assist:

- Trainee Scientific Glassblowers
- Those teaching Trainees
- Research Scientists and Chemists to gain a better understanding of what may or may not be possibly by outlining both the advantages and limitations of glass in the laboratory.
- Those wishing to gain an insight into glass manipulation

This manual is graded in small steps, each step is sufficiently small that the trainee will be able to master when given time to practice, which in turn will create both an understanding and also confidence.

It is important that each step is mastered before moving onto the next stage. I would advise that two trainees work together in unison, learning from each other's mistakes and successes.

Each action is described fully with the help of photographs. I believe that the description I have given will enable the trainee to progress.

There is so much to take in especially the unusual sensation of having a material change from solid to liquid and back again supported between ones hands, this together with having a flame so close raging as well over 2000°C (two thousand degrees centigrade). This is a great deal to take in at first and can be overwhelming.

Therefore I would advise that each trainee read the section prior to starting. Trainee one will practice the task and Trainee two will read out loud each action to be taken.

The advantages of seeing from a different perspective the basic manipulation benefits both trainees especially when they reverse roles and manipulate the glass.

Having a second person repeat out loud each stage reassures the Trainee quickly and avoids the need to stop and read themselves. This will in some way be as if a fully Trained Glassblower is close by watching the trainees every move.

One will be learning how to coordinate both the left and right hand and given time each hand will be working in unison. Each small stage will flow together just as the very glass flows together.

As in music it is the gaps in time between the notes that create the tune, likewise in glassblowing, at first each action has to be practised before being mastered then blended together, it is the pauses as the glass surface cools ever so slightly that are reflected in the final piece.

Be aware that in the main that Scientific Glassware will not be blown into a mould but into "free air" which will require the Glassblower to control his/her actions in rotating the glass constantly to overcome the effects of gravity on the softened glass and to pause to allow the outer surface of the molten glass to cool slightly before finishing with a controlled blow against the cooling glass.

I have chosen not to wear my white coat for the black and white photographs in the manual purely to help define the clear glass. The colour plates are spectacular especially where they show the residual strain under polarised light. During the past fifty years I have always kept a photographic record or sample of work that was of interest to me as a glassblower as it "made me think".

Some glass by its very nature is beautiful, occasionally a piece of glass can go beyond functionality in enabling the end user to proceed with a project, their appreciation of the glassware has made me feel very humbled to have used my accumulated skills to be part of their work.

In the final section of this manual I have drawn upon the experience of respected glassblowers to describe a particular item of glass that they have been involved with. I have asked that they describe not only the bending of the glass but the whole thought process from consulting with the scientist or engineer to the final item including any problems encountered along the way and the manner in which they were dealt with.

The whole philosophy of Scientific Glassblowing is that one works from the "inside — out" with a material that changes phases from solid to liquid then returning once more to solid. Its shape often defined by the stress levels within, while all the time being aware of the constant effect of gravity.

In Section 5.17.9 Button Seal alternative methods I have described how to make a simple water vacuum pump. I have included a photograph of one I had made while undergoing my initial training in 1966. It was made in soda lime glass and I was very proud of it and took it home to show my rather bemused mother which is probably why it has survived.

When I was asked if I would be prepared to write this manual I thought for a moment of those who gave their time to train me and set the standards to which I was to aim in the quality of my work and the support and respect that I should show to those who would use my work.

The people who taught me are now all gone and so has all of their accumulated knowledge. I knew that I owe them so much and feel that I have to record and pass on in some small way the skills that over time have enabled various branches of science to flourish.

I make no apology in recording their names:

Alfred Behrens, David Greenhalgh, Norman Collins, Cyril Blackburn, Richard Caveney, Terry Fallon, Wim Van de Bospoort. I offer my grateful thanks.

Paul Le Pinnet
November 2015

Contents

Acknowledgements		v
Foreword		vii
Introduction		xi

1	Glass Technical Data		1
2	Basic Tools and Equipment for Scientific Glassblowing		11
3	Working Environment Regulations — Workshop Layout		17
4	Safe Glass Handling		25
5	Hand Techniques		33
	5.1	Cutting of Glass — Tubing, Rod and Flat	33
	5.2	Spear Points	49
	5.3	Straight Joins	59
	5.4	Unequal Joins	66
	5.5	T Pieces	74
	5.6	Y Pieces	77
	5.7	90º Bends	81
	5.8	U Bends	91
	5.9	Button Seal	100
	5.10	Supported Internal Seal	110
	5.11	Items 1–9 Repeated in Progressive Larger Diameters	121
	5.12	Bulb Blowing	149
	5.13	Spiral Winding	161

5.14	Liebig Condenser	167
5.15	Spiral Condenser	175
5.16	Capillary Joins	186
5.17	Repeat — Alternative Techniques	196
5.18	Cones, Sockets, Spherical Joints, Screw Joints, Flat Flanges and Buttress Joints — Their Application — The Use of Mechanical Aids Such as Lathes, Cut-Off Machines, Linisher Belts Diamond and Carborundum Lapping Machines	211
5.19	Flange Making — Dewar Seals, Sinters	236
5.20	Safe Working Pressures (Positive) — Vacuum	250
5.21	Glass to Metal Seals, Recrystallised Alumina to Glass Seals Graded Seals	254
5.22	Various Techniques	271
6	Hand Torch Techniques	311
7	Vacuum Manifold — Vacuum Measurement — Schlenk Lines	335
8	Silvering of Glassware	347
9	Packed Columns	353
10	Round Bottomed Flasks — Various Techniques for Attaching Side Arms	357
11	Standards of Competence	367
12	Drawing on Experience	371
12.1	The Manufacture of a Quartz UV High Temperature Coil Photoreactor	371
12.2	Laser Cutting Quartz Glass	374
12.3	Fit, Form and Function	376
12.4	Combining Artistic Glass Working Techniques within a Scientific Glassware Context	379
12.5	The Foster Cell	389
12.6	The Design and Construction of his Masterpiece	398
12.7	Variable Temperature Gas Inlet for Surface Science Ultra-High Vacuum Molecular Beam System	408
12.8	Former Apprentice of Seven Years	420

Index   433

1
Glass Technical Data

The figures should be used for reference only. Data for design and product application should be taken from the respective Company's technical data.

Working Temperature

This is the temperature at which the glass may be fully reshaped and will sag under its own weight.

Softening Point

This is the temperature at which unsupported glass will begin to sag.

Annealing Point

This is the temperature at which glass will relieve stresses in a matter of minutes. To anneal glass, whether during manufacture or after processing all parts of the glass must be cooled uniformly from over the annealing point to under the strain point.

Strain Point

The extreme upper limit of serviceability for annealed glass and above which permanent stress can be induced into a glass. The maximum "inservice" temperature is always below this point.

Coefficient of Thermal Expansion

The property of a material to expand in size as the temperature is raised. Most glasses have a relatively linear expansion rate between 0°C and 300°C.

The lower the expansion, the greater the resistance of a glass to sudden temperature changes. Glass with an expansion coefficient of 33×10^{-7}°C would mean that a one metre long rod will expand 0.33 mm when heated an additional 100°C.

7740 Corning Pyrex

Thermal expansion	0–300°C = 32.5×10^{-7}°C
Strain point	510°C
Annealing point	560°C
Softening point	821°C
Working point	1252°C

8330 Schott Duran

Thermal expansion	0–300°C = 33×10^{-7}°C
Strain point	525°C
Annealing point	560°C
Softening point	820°C
Working point	1260°C

KG-33 Kimble KIMAX

Thermal expansion	0–300°C = 32×10^{-7}°C
Strain point	513°C
Annealing point	565°C
Softening point	827°C
Working point	1255°C

Kavalier Simax

Thermal expansion 0–300°C	= 33 × 10^{-7}°C
Strain point	510°C
Annealing point	560°C
Softening point	820°C
Working point	1260°C

These hard borosilicate glasses "3.3" are characterised by high thermal stability as defined by the international standard ISO 3585: Borosilicate glass 3.3 — Properties (or by the identical Czech version of the standard NISO3585: Borosilicate glass — Properties or German DIN ISO 3585).

The glass fully meets the requirements set down by these standards.

Pyrex 7740 and Duran 8330 belong to this same class.

Soda-lime Glass

Thermal expansion 0–300°C	= 93.5 × 10^{-7}°C
Strain point	473°C
Annealing point	514°C
Softening point	696°C
Working point	1005°C

Fused Silica

Thermal expansion 0–300°C	= 5.0 × 10^{-7}°C
Strain point	890°C
Annealing point	1020°C
Softening point	1530°C
Working point	2000°C

Annealing

Annealing is a thermal process whereby the strain induced in the glass from forming or working is removed.

The annealing cycle consists of three stages in an annealing oven.

The temperature increase at a controlled rate (ramp) will depend on the thickness of the glass and mass of the material in the oven.

Dwell time at the annealing temperature is to allow all strain induced in the glass to dissipate and will vary with the wall thickness.

Cool down from the dwell, then from the lower annealing temperature back to room temperature.

After annealing it is advisable that the glassware is viewed under polarised light to ensure that there is no residual strain and the glassware is fully annealed. This in effect is a check on the oven itself.

I tend to set the annealing oven a few degrees above the glass annealing temperature and allow the oven to "cut off" on reaching temperature. The residual heat within the brickwork of the oven is quite sufficient to anneal the glass as the Dwell time for glass with a 3-mm wall thickness is 5 minutes.

Glassware with much greater wall thickness will require longer Dwell times.

Wall thickness	Dwell time
3 mm	5 minutes
6 mm	10 minutes
9 mm	18 minutes
12 mm	30 minutes

Dwell times for the thicker walled glass must be held at or maintained at the times indicated noting that these are minimum Dwell times.

Annealing ovens are brick lined and therefore there is a great amount of residual heat which will dissipate over time. Be prepared to allow the oven to cool down overnight although sending the oven up during the day is often required but I would advise not opening the oven door until the temperature has dropped below 400°C.

Larger scale production line glass manufacturers use an annealing Lehr. The glassware being placed on a moving bed which traverses the glassware through the Lehr at a controlled speed past various sections of Lehr which are temperature controlled to ramp up the heat, dwell and cool in a continuous system.

Glass annealing Lehrs are suitable only for large-scale manufacture based upon their size alone and energy usage. The speed at which the glass passes through the annealing process is far quicker than the batch process to be found in a workshop although there are various ovens to be found such as fan assisted, small 30 × 30 cm² for quicker turn round and larger 75 × 67 × 208 cm. All is dependent upon the type and size of the glassware to be annealed regularly.

When large-scale Chemical Plant glassware is being produced, large annealing ovens on wheels are "brought" to the glass rather than vice versa.

The following list of glasses and their applications are for reference. They will show the wide range of applications of technical glasses available.

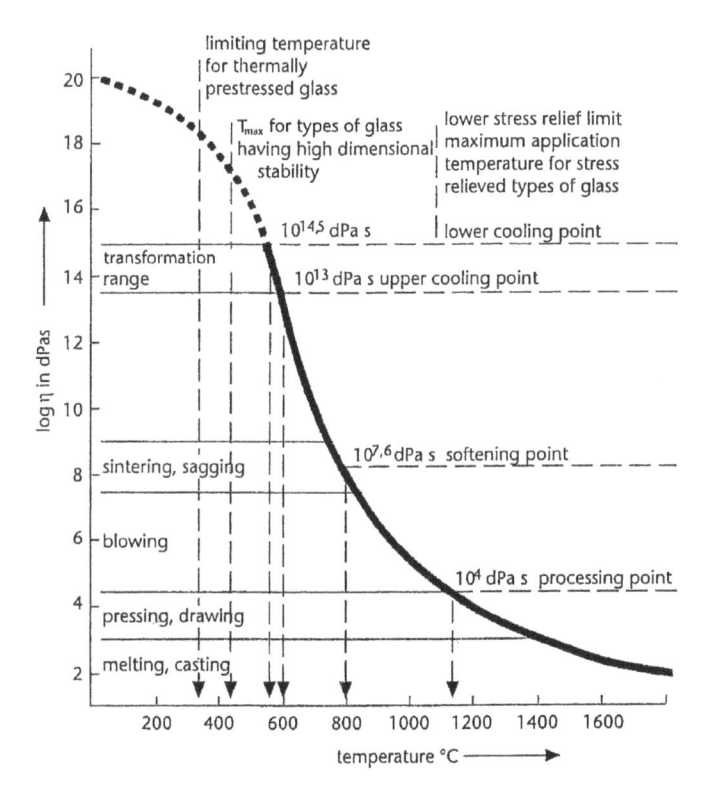

Fig. 1.1 Normal temperature dependence/viscosity curve for DURAN®. Viscosity ranges of important processing techniques, the position of fixed points of viscosity and various limiting temperatures.

The following information was supplied by SCHOTT AG, Germany, and uses their reference numbers.

8095 — Lead glass 28% P60, electrically highly insulating for general electro-technical applications.

8100 — Lead glass 33% P60, electrically highly insulating, highly X-ray absorbing.

8228 — Intermediate sealing glass.

8229 — Intermediate sealing glass.

8230 — Intermediate sealing glass.

8240 — Alkaline earth aluminosilicate glass for high temperature application in electrical engineering for sealing molybdenum free from alkali, blue coloured, defined absorption.

8241 — Alkaline earth aluminosilicate glass for high temperature application in electrical engineering.

8242 — Borosilicate glass for Fe–Ni–Co alloys and molybdenum, electrically highly insulating.

8245 — Sealing glass for Fe–Ni–Co alloys and molybdenum, minimum X-ray absorption, chemically highly resistant.

8250 — Sealing glass for Fe–Ni–Co alloys and molybdenum, electrically highly insulating.

8252 — Alkaline earth aluminosilicate glass for high temperature applications for sealing to molybdenum.

8253 — Alkaline earth aluminosilicate glass for high temperature applications.

NEO 1730 — Alkaline earth neodymium containing violet colour aluminosilicate glass for high temperature applications in electrical engineering for sealing molybdenum free from alkali.

8270 — Borosilicate glass for sealing to KOVAR metal and molybdenum electrically highly insulating defined UV absorption, stabilised against solarisation.

8326 — SBW glass, neutral glass tubing, chemically highly resistant.

8330 — DURAN®, BOROFLOAT®, and SUPREMAX® borosilicate glass, all-purpose glass mainly for technical applications such as apparatus for the chemical industry, pipelines and laboratories (see Fig. 1.1). The figure shows the relationship between viscosity and temperature for DURAN borosilicate type glass, a higher working temperature being required for casting, pressing, and drawing whereas blowing is somewhat lower and sintering and sagging (deformation) are shown to be lower still.

8337B — Borosilicate glass, highly UV-transmitting, for sealing to glasses and metals of the KOVAR/VACON — 10/11 range and tungsten.

8341 — BOROFLOAT®40, borosilicate float glass adapted for thermal toughening.

8347 — Colourless, highly TR8330.

8350 — AR-GLAS®, soda-lime silicate glass.

8360 — Soft glass, lead-free.

8405 — Highly UV-transmitting soft glass.

8412 — FIOLAX®, clear neutral glass, chemically highly resistant for pharmaceutical packaging.

8414 — FIOLAX® Amber, neutral glass, chemically highly resistant, for pharmaceutical packaging.

8415 — ILLAX®, amber glass for pharmaceutical packaging.

8436 — Glass, particularly resistant to sodium vapours and alkaline solutions, suitable for sealing to sapphire.

8447 — Intermediate sealing glass.

8465 — Low melting lead–alumina–borosilicate solder glass. Good match with materials with thermal expansion 8·5–9·5 ppm/k.

8470 — Lead-free borosilicate solder glass for sealing of materials with thermal expansions of 10·5–11·5 ppm/k.

8487 — Sealing glass to tungsten.

8488 — SUPRAX®, borosilicate glass, chemically and thermally resistant.

8516 IR — absorbing sealing glass for Fe–Ni, lead-free, low-evaporating.

8531 — Soft glass, sodium-free, high lead content for encapsulation of semiconductor components at low temperatures (diodes).

8532 — Soft glass, sodium-free, highly lead-containing, for encapsulation of semiconductor components at low temperatures (diodes).

8625 — IR-absorbing biocompatible glass for (implantation) transponders.

8650 — Alkali-free sealing glass for molybdenum, especially for implosion, diodes, highly lead-containing passivation glass.

8651 — Tungsten sealing glass for power and PIN diodes, passivation glass.

8652 — Tungsten sealing glass for power and PIN diodes, low melting passivation glass.

8660 — Borosilicate glass for sealing to tungsten of high caesium content.

8689 — Borosilicate glass, highly UV-blocked, stabilised against solarisation, sealing glass to tungsten.

8708 — PYRAN® Platinum, transparent floated glass-ceramic.

8800 — Neutral glass, highly chemical resistant.

The following items are not available in tube form.

G017-002 — Composite passivation glass, consisting of zinc-borosilicate glass and inert ceramic filler, for passivation of semiconductors.

G017-004 — Lead-borosilicate glass with an inert ceramic filler for passivation of semiconductors.

G017-052 — Low melting lead-borate glass.

G017-096R — Lead-borosilicate glass for passivation of semiconductors.

G017-230 — Composite glass for passivation of semiconductors, especially power transistors.

G017-339 — Non-crystallising, very low melting composite solder glass, consisting of lead-borate glass and inert ceramic filler, good match to materials with a CTE of 5–6 ppm/k.

G017-340 — Non-crystallising, very low melting composite solder glass consisting of lead-borate glass and inert ceramic filler.

G017-388 — Passivation glass for thyristors and high block rectifiers. Also usable as solder glass.

G017-393 — Non-crystallising low melting composite solder glass with a lead-borate glass and inert ceramic filler. The thermal expansion coefficient is well matched to alumina.

G017-712 — Low melting composite solder glass consisting of a lead-borate-based glass and inert ceramic filler.

G017-725 — Lead-borosilicate glass for passivation of diodes.

G017-997 — Composite glass for passivation of silicon wafers based on a lead-silicate glass and an inert ceramic filler.

G018-133 — Passivation glass for sinter glass diodes.

G018-197 — Low melting, lead-free passivation glass.

G018-200 — Lead-free passivation and solder glass.

G018-205 — Zinc-borate glass suitable for hermetic capsulation of diodes and general sealing glass.

G018-228 — Lead-containing composite solder glass for CTE range 7–8.

G018-229 — Lead-containing composite solder glass for soda-lime glass.

G018-249 — Low melting lead-free solder glass.

G018-250 — Low melting lead-free solder glass for the thermal expansion range of 7–8.

G018-255 — Lead-free solder glass for CTE range 9–10.

G018-256 — Non-crystallising, low melting lead-borate glass.

G018-266 — High temperature sealing glass for Al_2O_3.

G018-281 — Glass-ceramic sealant for SOFC applications.

G018-311 — Alkaline-free high temperature sealing glass for ZrO_2 ceramics.

G018-339 — Alkaline-free high temperature sealing glass for ZrO_2 ceramics.

G018-346 — Alkaline-free high temperature sealing glass for Al_2O_3.

G018-354 — Glass-ceramic sealant for SOFC applications.

G018-358 — Alkaline-free high temperature sealing glass for Al_2O_3.

G018-381 — Glass-ceramic composite sealant for SOFC applications.

G018-385 — SOFC sealing glass/high temperature sealing glass for Al_2O_3.

GM3-1107 — Glass sealant for SOFC applications.

2

Basic Tools and Equipment for Scientific Glassblowing

*Information supplied by **Jeremy Bolton**
of Jepson Bolton & Co Ltd., Watford, UK*

Scientific glassblowing can be dangerous. In a studio, busy workshop or production facility, there are large and small naked flames as well as hot and cold glass everywhere so before setting up a new workspace, it is wise to spend a few moments considering location and safety of both the glassblower and potential visitors.

The basics you need to get started practising with tubular or solid rod borosilicate glass are as follows:

- Large Heavy Duty Workbench (approximately 2 m wide × 0.8 m deep) with suitable supportive seating.
- 2 or 3 Gases Bench Burner.
- Propane and Oxygen Gas Bottles with regulators, Flash Back Arrestors and hoses.
- Didymium Safety Glasses.
- A selection of basic glass working hand tools.
- A blow Swivel, hose and single hole stoppers.
- A wooden or metal cooling rack or rest.
- A large metal empty bin and a small bucket of water.
- Tungsten Glass scoring knife.

Workbench

The workbench can have a heat resistant top or to keep costs down some 10 mm thick ply sheeting which can cope with hot glass being rested or dropped on it. You need to be able to get your legs under the workbench to sit comfortably behind the bench burner.

Burners

The choice of bench burner will depend on the work you expect to undertake. There are two main types of burners — Premixed and Surface Mix. In the premixed burners, the combustion gases are mixed inside the burner body so when lit at the burner nozzle they are already mixed. In surface mix burners, the combustion gases are mixed on the surface of the burner head outside of the burner body. Premixed burners tend to be louder but give a more concentrated hotter flame. Surface mix burners have been gaining popularity due to their quietness, range of flame sizes, flexibility and stability.

A popular Surface Mix Model is the Arnold Zenit No. 114 bench Burner as per the photograph in page 12. This unit has a large flame range, is easy to control with one knob enabling you to go from a large bushy flame down to pin prick sized flames very quickly.

Gases

The choice of gases you use in your workshop will be determined by a number of factors such as what is easily available locally, if there is any mains gas available in the workshop building and the type and expected sizes of glass you wish to manipulate.

All the fuel gases are burnt with oxygen and each has its own advantages and disadvantages.

Natural gas has a BTU value of 1012 and has the benefit of being a clean burning gas although it does not achieve as higher flame temperature as the other two gases. If available in your workshop building, it is normally the cheaper option long term. It can also be burnt with just compressed air giving a lower temperature flame for certain applications.

Propane or butane gas that has a BTU value of 2557 gives a higher flame temperature and has the benefit of being readily available in large bottles to give convenience and portability. It is not as clean burning and the flame with oxygen gives off a bright glare which can be reduced by adding a small amount of compressed air to the burner.

Hydrogen that has a BTU value of 324.2 gives the highest flame temperature and is normally the most expensive of the gases to buy. It is mostly used by glassblowers who process quartz glass. The flame is extremely hot, very clean, almost invisible. This gas is used when the possibility of any impurities touching the glass during production need to be minimised.

Didymium Safety Glasses

When borosilicate glass is heated in a flame, it develops an amber glow around it which obscures the hot glass area. The use of Didymium Spectacles filters this sodium glare and allows the wearer to see their workpiece clearly. One of the latest developments is to incorporate an additional filter on the lenses to remove UV reflective rays from the workpiece. There are many different styles of these glasses available.

Glass Working Tools

The basic tools for bench glass manipulation were traditionally made of brass and used with bees wax to prevent the brass from sticking to and marking the glass. Nowadays, the most popular tools are made from carbon and are available in many different sizes and shapes. Molten glass can be shaped and manipulated easily without leaving any residue.

The tools you require will depend on the size and type of glass manipulation you wish to undertake. Below is a list of a starter set of tools for bench work.

Stainless Steel Tweezers 6" and 8" Long.
Carbon Paddles 75 mm × 50 mm and 150 mm × 100 mm with handle.
Carbon Plate 200 mm × 100 mm × 10 mm thick.
Carbon Rods, various sizes 3 mm, 5 mm, 7 mm, 10 mm and 15 mm.
Carbon Reamer in triangle shape 75 mm × 50 mm.
Selection of cork stoppers to match tubing sizes.

The possibilities of manipulating and forming glass for scientific, industrial or artistic applications are endless and the ease with which you can adapt and create tools to help you achieve what you wish is relatively straightforward. Carbon can be machined or cut to make tools or shapes for forming and be made into moulds for blow forming.

There is also a large range of tooling accessories for gripping and holding all types and shapes of glass for all the different stages of piece production.

Advanced Machinery for Processing Larger Glass Tubes and Rods

Glassblowing lathes are available from small 25 mm bore desktop versions through to floor mounted lathes which have a bore of 1000 mm. They can be manufactured for manual, semi or fully automatic operation. Many hi-tech products we use today such as fibre optics and electronic chips have been processed through glass products which at one stage of production were made with these types of machines.

All glassblowing lathes consist of a solid bed on which the headstock, tailstock and burner carriage are mounted. The twin ground and hardened guide bars ensure the two stocks are precisely aligned and gives easy movement along the guide bars. There are adjustable chucks on each stock which are to be adjusted to grip the glass firmly.

The photo shows a 60-mm Bore Bench Lathe Model 1060 with two Scroll Chucks.

The control panel allows the operator to adjust the speed and direction of rotation. There is an emergency stop and an electronic brake function which locks the chucks in one position for ease of opening and closing. There is also the ability to disconnect the tailstock from the drive shaft. This then enables the tailstock to be used to hold tooling for forming operations on glass being held in the headstock.

There is a drive shaft running along the back of the lathe which drives the rotation of the chucks synchronously.

The burner carriage has T-bar slots on it so that many different types of burners or forming tools can be mounted. The latest innovation has seen standard gas burners replaced with lasers so that an even finer heating zone can be processed for the most demanding of technical sealing and joining applications.

The glassblowing lathe is deceptively simple but it has to be able to operate under extreme temperature ranges and for long periods of time and still maintain its alignment and accuracy and free easy movement.

Many different tooling stations and accessories can be added to the basic lathe to enable the processing of hundreds of glass forming, sealing and joining operations.

Further Equipment Used in the Scientific Glassblowing Workshop

- Glass cutting machinery;
- Annealing ovens;
- Vacuum systems;
- Grinding machines;
- Drilling machines.

3

Working Environment Regulations — Workshop Layout

Be aware that there are both national and local statutory bodies that set and monitor all regulations covering the health and safety standards within the workplace.

One must adhere to and be aware of each regulation covering new material handling and storage, working conditions and personal protection equipment through to the final product which should be fit for purpose. Regulations are in place meaning that one must work within boundaries for one's own safety and the safety of others.

Oxygen cylinders are both heavy and relatively unstable due to the small diameter of the base in relation to their height; store cylinders vertically in a well ventilated area, securely chained or bracketed to a substantial wall or post. Always segregate empty and full cylinders. Refrigerated liquid oxygen containers should also be sited in a well ventilated area securely within a gated concrete bunded area while being surrounded by substantial fencing of sufficient strength to shield against accidental collision.

The oxygen supply to the workshop must have cut off and control valves together with flash back arrestors and a pressure relief valve fitted permanently within the system.

All equipment must be free from oil or grease, reducing agents, combustible materials and organic materials.

All control valves and pressure relief valves must be routinely tested by an accredited engineer; control regulators must be replaced and refurbished in accordance with national and local regulations.

Oxygen Enrichment — The Risks

Normal air contains:

21% Oxygen, 78% Nitrogen, 1% Argon inc. traces

Any increase in percentage of oxygen has an effect on life and the combustion process, oxygen enriched atmosphere even by a few percent increases the risk of fire considerably.

When an oxygen pipeline enters a building, an isolation valve should always be provided on the outside of the building.

Flammable fuels in cylinders can be used within the workshop after being fitted with the appropriate control regulators and flash back arrestors.

Larger volumes must be stored outside within a protected area — the distance from other buildings and flammable materials should be closely regulated by local and national bodies which must be adhered to for the sake of both employees and local residents.

Pipework carrying either oxygen or fuel gasses must be sound, robust and rigid to the point of use where at which point flexible connection to the burners may be made.

Fuel gasses have an asphyxiation hazard but they also present the greater hazard of fire and explosion.

If a gas container is kept inside or in a confined space, even small quantities of escaping fuel gas are sufficient under certain conditions to form an ignitable mixture.

There is only a small risk that the lower limits will be reached in workshops that have good ventilation.

For this reason many fuel gasses are "stenched" — given a smell like rotting fish which means that leaks are readily identified.

Extraction Systems

- Are either "overall" sweeping the entire workshop which requires a minimum of 20 changes of air per hour without causing draughts which may affect flame shape at the bench or lathe.

- Or "local" where a flexible extracts tube is placed close or directly over the work. This system is often used when working with fused silica as this uses very high temperature which may cause silica fumes.

The extract system must be thoroughly tested on a regular basis: annually.

The incoming air can be directed through filters which in turn should be renewed, twice per year.

If available, waste steam can be used to prewarm the incoming air — not essential but very much appreciated during winter months.

With such high temperatures being generated by oxygen/gas burners the production of Carbon dioxide is high as is nitrous oxide and nitrogen dioxide.

Nitrogen Dioxide

The World Health Organization has described as "Chronic exposure" concentrations of 20 parts per million (ppm) with immediate danger to life.

If there is a high concentration of nitrogen dioxide it may replace oxygen and cause asphyxiation.

Nitrogen dioxide may increase problems for asthmatics in concentrations as low as 1 part per million (ppm). 5 ppm nitrogen dioxide can cause bronchitis, emphysema, respiratory irritation.

Nitric oxide is a short-lived pollutant which is converted to nitrogen dioxide in the presence of free oxygen.

All of which is sufficient information to ensure that when working hot glass either at the bench or lathe have the extract system SWITCHED ON!

The following overview for electrical installations found within the workshop are to be borne in mind with respect to new installations and established electrical equipment. The regulations change over time therefore the regulatory authorities should be contacted and advice should be sought to ensure conformance.

All electrical cables to be protected within suitable enclosures, for example, steel conduit or trunking.

All electrical sockets to be protected by residual current devices (RCD) or residual current breaker with overcurrent (RCBO) systems.

Lighting

When rotating machinery is illuminated by pulsed neon tubes, there is the possibility that the "neon pulse" may cause a stroboscopic effect which gives the illusion that the rotating machinery is "Static". When illuminating an area with lathes, incandescent lighting alleviates any possibility of the stroboscopic effect taking place.

All electrical installations should be tested every three to five years and portable appliances tested (PAT) annually.

Fire alarms should be tested weekly and all exit procedures annually.

Exit signage must be appropriate for both workforce and visitors who should be escorted to safe areas and assembly points.

New employees must be made aware of the emergency shut down procedures; the procedures should be under constant appraisal.

Emergency lighting should give a minimum of 1 lux (moonlight) and should guide people to the nearest exits for at least three hours.

Fire alarm sensors should be compatible with the environment in which they are to be used.

Glassblowing workshops with naked flames compared with an office environment have visibly different requirements. The type of sensor must be selected accordingly.

The high frequency and level of noise generated by certain torches should be monitored — and the long-term effect on the user also. Certain burners may require the use of ear defenders — signage to that effect must be obvious and close to the work area.

Staff must be made aware of any safety concerns and as a consequence given training if necessary. Furthermore, all trainees must have supervised training from accredited competent people when first using equipment such as: cut-off machines, drilling machines, lathes, grinding machines, edge finishing machines, linishers, lapping machines to name but a few.

Formal training must be given when subject to national legislation, such as use of grinding wheels.

Lifting equipment when available to replace heavy chucks from lathes must be checked annually by qualified personnel.

The overall layout of the workshop must take into consideration both the storage of the raw material and its passage through the workshop.

Store tubing and rod horizontally clearly marked and located in such a way as not to interfere with normal walkways (see Fig. 3.3).

When 1·5 m lengths of glass tubing are being removed from their storage, the glass or the person should not be in a position to compromise any nearby burner flames or cause a distraction; neither should the movement through the workshop. The glass should be carried vertically and close to the body and not wafted randomly about in the air.

Trainees in particular must be taught by example, that glass is not a material to be "played with" in a Scientific Glassblowing workshop.

Housekeeping can deteriorate as the pressure of work increases. This happens but it is not an excuse — floor areas, thoroughfares, work areas such as benches and lathes, cold working areas. All have a collective responsibility to maintain a standard of housekeeping agreed and understood by everyone, a copy of the "understanding" should be displayed prominently. This will not then be an imposed set of rules.

The disposal of waste glass, chemicals and solvents are governed closely by statute of which one must be aware.

Personal Protection Equipment (PPE) must be available to both staff and visitors. Especially eye protection for visitors, which must be considered the highest priority. Eye protection for glassblowers to remove the sodium flare which may mask the molten glass would consist of didymium lenses, either prescription or clip on. There is a greater need for this type of protection when using soda-lime glass as the flare is quite intense (Fig. 3.1). The sodium flame becomes less obvious in a well-illuminated work area. Didymium lenses remove the sodium flare although some glassblowers prefer to use CRUX B2 lenses which shade out most of the flare but allow one to see part of the sodium flare so as to gauge the prewarm more accurately (Fig. 3.2).

When working with fused silica it is essential that dark glasses are used as the glass when hot creates an intense white light which is damaging to the eyes.

One may choose to wear laboratory coats to reduce the amount of radiated heat from the work while remembering that white will reflect and dark colours will absorb the heat — your choice.

Heat shields when melting large diameter silica are an essential element of high temperature work.

22 *Laboratory Scientific Glassblowing: A Practical Training Method*

Fig. 3.1.

Fig. 3.2.

Standard scientific glassblowing needs a good supply of heat resistant gloves when transferring hot glass to the annealing oven or when adjusting hot glass when in the lathe. There are numerous needs and opportunities for good effective hand protection.

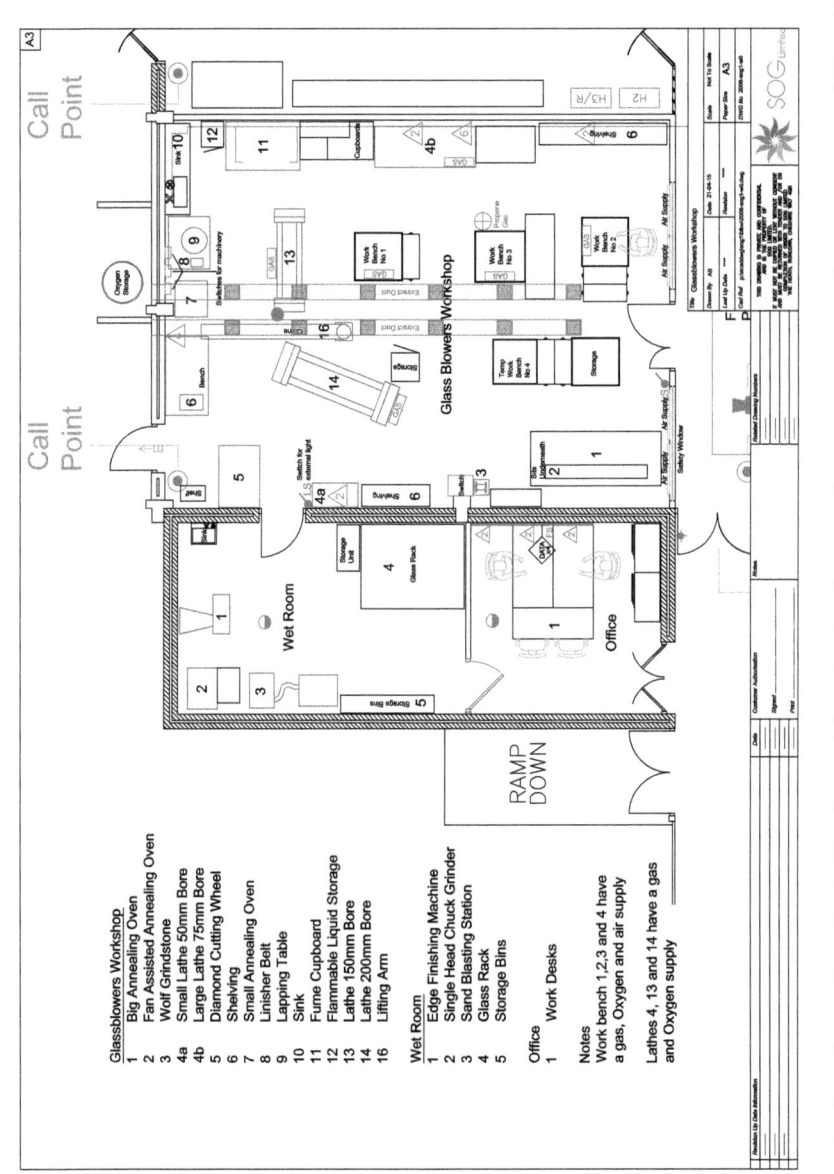

Working Environment Regulations — Workshop Layout 23

Fig. 3.3. A schematic overview of a glassblowing workshop showing the external oxygen storage, central extracts system and filtered input.

I have taken to marking my gloves with the letters L and R in bold letters not because I am "simple" but for the simple reason that on occasions heat resistant gloves are needed immediately! They all look alike. I seem to have far more left hand than right and this simple expedient of marking the gloves has saved work from failure as I have been able to quickly pick the correct glove without hesitation.

In conclusion, one may be aggrieved at a list of do's and don'ts when faced with health and safety issues. Bear in mind that most safety legislation has evolved over time (and is still evolving) due to the experience of others, the consequences in some cases being injury or death.

A simple example relates to poly-tetra-fluoro-ethane (PTFE), a material in common use, especially in terms of glassblowing which has tended to move away from greased taps and now uses PTFE keys within a polished barrel. Equipment for repair may have PTFE keys securely embedded in the barrel!

Remove the PTFE key or piston without fail before working on the repair or even annealing. WHY?

PTFE when subjected to temperature above 300°C degrades to form

Perfluroisobutylene fumes which are extremely toxic. The fluorine element will etch and degrade the glass making it unusable.

That is experience — learn from it.

Also, the very important emergency exits from both the workshop and the associated office area.

4

Safe Glass Handling

The aim of the Scientific Glassblower is to "enable" those who are laboratory-based to conduct small-scale chemical reactions using equipment that is transparent which enables one to see a colour change for example, or crystals coming out of solution, liquid–liquid separation etc. The observation of change indicates that some form of chemical reaction has taken place.

The glass in common use within the laboratory is a Borosilicate glass which is relatively inert although the glass does not react well with hot phosphoric acid, hydrofluoric acid or alkali.

Borosilicate glass has a safe working pressure as indicated in the adjoining graph. The graph indicates maximum working pressure for Borosilicate glass of various wall thickness and diameters. The graph assumes the glass is freely supported and not subject to clamping forces and does not apply to cylindrical tubes with flat bottoms (Fig. 4.1).

Where glass has been "worked" in any way such as side arms or jacketed vessels, a maximum safe working pressure of two atmospheres should be taken.

The whole point of this safe glass handling section is for both the Glassblower and the end user to understand both the limits and advantages of using glass in the laboratory. It is essential that the Scientific Glassblower understands the material and asks relevant questions of the Scientist regarding under what conditions the glass is to be used.

The Scientific Glassblower will at times be asked for advice by those who may well be new to laboratory work. The safety of the end user may well be dependent upon the advice given.

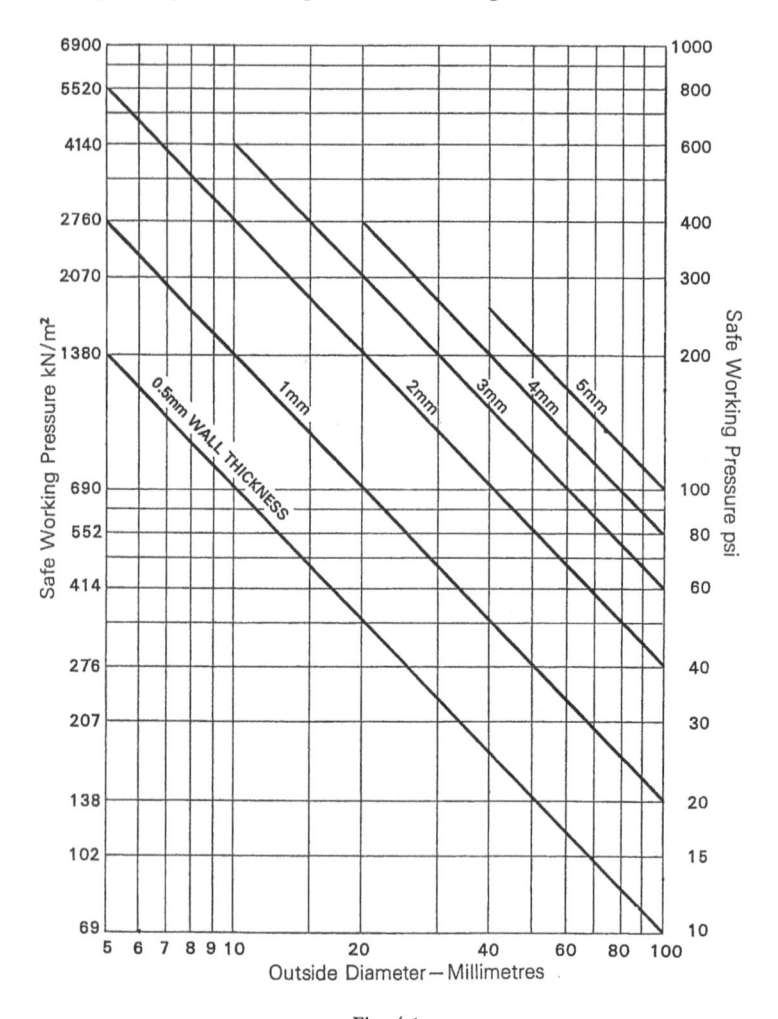

26 *Laboratory Scientific Glassblowing: A Practical Training Method*

Fig. 4.1.

Glass is a versatile material in that it can contain, deliver and by jacketing reaction flasks control the speed of reaction by either cooling or heating, while under vacuum or positive pressure.

Laboratory scale chemical reactions can confirm or otherwise the feasibility of a chemical process which in turn may be scaled up to pilot plant level before committing to full scale manufacture of chemical plant.

Precision delivery of liquids via burette or pipette and the world of optics are to a great extent still glass-based.

Glass is an inorganic super cooled liquid. It is non-crystalline.

In Glassblowing terms, it is a stiff liquid at room temperate which when heated in a flame will soften and flow like a liquid. It does not have a specific melting point but a softening point of 821°C and a working point of 1252°C for borosilicate glass. Soda glass being 696°C melting and 1005°C working point soda glass is used less and less in laboratories because of its susceptibility to thermal shock. Fused silica with a melting point of 2000°C is used for specific high temperature work.

The maximum temperature to which Borosilicate glass can be subjected to in a laboratory is 500°C for short periods only, bearing in mind that the temperature at which strain is induced into Borosilicate glass is 510°C.

However, once the temperature exceeds 150°C, care should be taken that the heating and cooling are achieved in a slow and uniform manner.

Glassware above 150°C should not be placed upon a cool surface as the thermal shock to the glass may cause the glass to fail.

When stirring solutions in glass vessels such as beakers and flasks, avoid using stirring rods with sharp edges which can scratch and weaken the glassware. Scratch marks are equally dangerous on the inside as on the outside of a vessel, most of the strength of glass is in its surface. Its strength will be reduced considerably especially if in contact with liquids together with either heat, pressure or vacuum failure *will* result. The consequences or using glassware with cracks or scratches however small they may be can be catastrophic if failure takes place in use, as the shards of glass and the contents make failure doubly dangerous.

Glassware in the laboratory is often collective in use. So it is essential that the glass is cleaned thoroughly after use as the user will have known its contents and is responsible for its cleanliness before putting away. The way we store glassware is important as glass on glass contact can cause the beginning of a "star crack" (Fig. 4.2) at the point on contact, round bottomed flasks are particularly vulnerable, so prevent glassware from touching by lining drawers and cupboards with cotton wool, bubble wrap or a similar material. Store large heavy items of glassware at ground level not on shelves, neither place large heavy items of glassware on top of delicate glassware — the results can be dramatic and dangerous. Storing beakers one inside another can cause the pouring spout to wedge on the inside of the larger beaker and cause injury to anyone trying to release them.

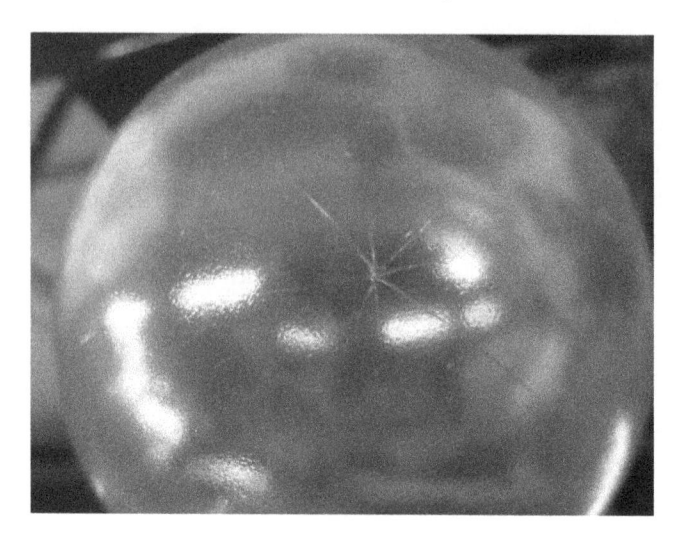

28 *Laboratory Scientific Glassblowing: A Practical Training Method*

Fig. 4.2.

Once ready to assemble the glassware, support the glassware from underneath either in a cork ring, heating mantle retort ring or oil bath but make sure that clamps are not misaligned and not overtightened.

Glass is strong in compression and weak in tension so "always" support from underneath and do not allow glassware and its contents to bare the weight without support.

One may need to make a tubing connection to the likes of condensers or Dreshel heads which are now found to be fitted with screw connections although one may find the need to make a connection directly to a glass tube, it is important that excessive force is not used and that the join is well lubricated with water while holding the glassware no further back than three diameters (Fig. 4.3), walk the tubing onto the glass then once more move the hand back a further three diameters.

The same procedure is used when attaching a bulb to a pipette. This will avoid putting too much pressure on the glassware.

The consequences of glass being mishandled and breaking may cause a cut to the flesh by a material that is sharper than steel which can slice through the cells of the flesh it may well inject whatever it had been containing directly into the bloodstream.

A point to note is that should one have to seek medical attention, the medical staff will be concerned that no broken pieces of glass remain in the wound will have to probe the wound to confirm all is clear. Bear this in mind

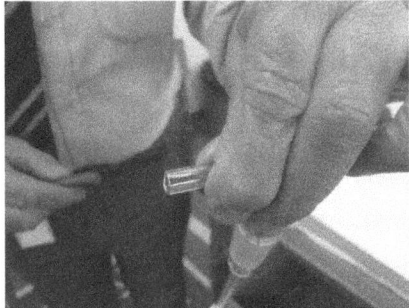

Fig. 4.3.

at all times when handling glassware, work in a controlled methodical manner, have your awareness raised at all times and work within all safety procedures.

When moving glassware around the laboratory, support the glassware with both hands at all times and when taking outside the laboratory use an approved carrier.

If glassware breaks in the laboratory, use thick PVC gloves to pick up the larger pieces and sweep up a much larger area than one would imagine as the smaller pieces can "fly" considerable distances.

How we dispose of broken glassware is equally important, if it cannot be decontaminated it must go through an approved recycling system for contaminated glassware and not just put in a bin or plastic bag. If one is sending glassware for repair, place it in an open labelled container so that it will not be used again.

One expects glass to be clear and bright in appearance, if the glass seems dull it may well have undergone chemical attack (Figs. 4.4–4.6), do not use it.

Do not put glassware into a container not designed for it such as a lab coat pocket.

Remember there is a fine line between genius and insanity — drinking out of laboratory glassware i.e. Beakers is insanity! Don't do it.

I have found that any documentation relating to safety can sometimes be viewed in a negative manner and regarded as just a list of do's and don'ts.

The very nature of working within a laboratory environment means that it is most important to raise and maintain safety awareness in oneself and the safety of others.

I have often been asked to run courses on "safe glass handling" and am very much aware that each laboratory differs in respect to the volume of

30 *Laboratory Scientific Glassblowing: A Practical Training Method*

Fig. 4.4. Hydrofluoric acid (minutes).

Fig. 4.5. Alkaline solution (days).

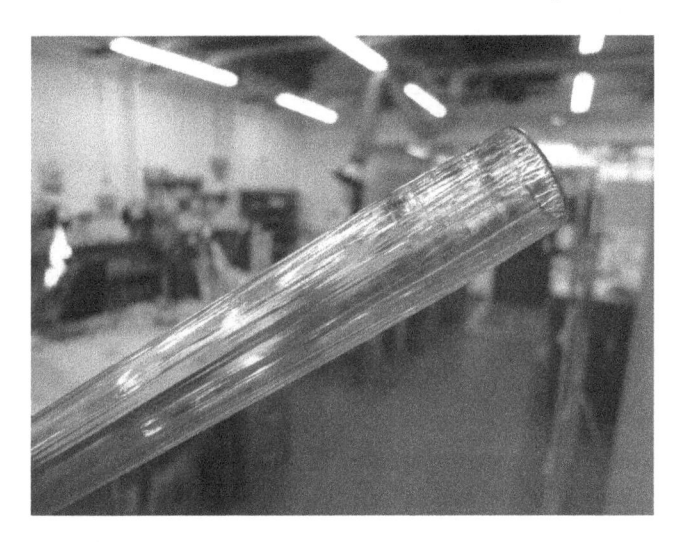

Fig. 4.6. Steam/water (months).

glassware being used, as a consequence a chemist with minimum contact with the material undergoing a career change or move can mean being confronted with a vast array of glassware, which has led one company Innospec Ltd of Ellesmere Port, Cheshire, UK to make the decision that all new starters working within their laboratories, no matter how much chemistry experience they may possess undergo a "safe glass handling" awareness course as part of their induction process.

Recently after returning to the Company some weeks after conducting such a course, I was delighted to see the following "Top 10 Rules for Glassware" which I was informed was drawn up by the chemists themselves without prompting.

It made me smile to see that they had framed the list which was then placed in all of the "wash rooms" on site, both male and female.

They owned the list.

Top 10 Rules for Glassware

- Keep glassware to a minimum, only have out what you will need. Think carefully about where you put it — both whilst in use balanced on bench edges and rolling about in drawers.

- Take care when handling glass — you know for a fact you can knock it off the bench and it will be fine but the moment you pick it up it will shatter into a thousand pieces.
- Glassware is a tool of your trade — treat it with respect and care you would any other tool — use the right piece for the activity — don't make do and mend.
- Check all glassware before you use it — look for cracks, chips or residual chemicals. If it's damaged or dirty, dispose of it safely and get a new one.
- Don't clean up broken glass with fingers — use a stiff brush (something that won't hold broken glass) and a shovel.
- We have a recycling stream for glass and chemical storage glassware within R&T (how good are we?). Don't put glass in the normal waste bins or skips — think of those who will have to handle it after you, the cleaners for example.
- It may be really obvious but the glassware you are using may get hot, cold or slippery when you are venting. Use thermal gloves or tongs to pick it up or move it around.
- Handling broken glass can introduce dirt and chemicals into your blood stream as well as cutting those silky soft hands — always use gloves or tongs.
- Lubrication is the key when inserting glass into rubber tubing
- There is a fine line between genius and insanity — drinking out of lab glassware is insanity (as well as unsanitary). Don't do it.

5
Hand Techniques

5.1 Cutting of Glass — Tubing, Rod and Flat

This section will cover basic hand techniques. The use of diamond wheel flame and laser cutting will be covered in Section 18.

Tubing and rod are supplied in 1500 mm length.

Carefully select a length of 8 mm ø tubing noting that the ends are flame polished (Fig. 5.1) during the manufacturing process. It is good practice for the student to similarly flame polish the ends of any tubing they may cut. Flame polishing will be described later in the text.

Place the tubing on a bench and mark at 300 mm or 12″ intervals with a fibre pen.

Grasp the tungsten carbide glass cutter in the right hand with the blade uppermost (Fig. 5.2), rest the glass on the blade at the marked position with the longer length of tube to the left (Fig. 5.3).

Place the thumb on top of the glass — grip the longer length in the left hand just to the left of the knife.

Rotate the glass tube towards you with the left hand while pressing down with the right (Fig. 5.4).

As the mark is made upon the glass there will be a sound of ripping silk.

There is no need to make a mark around the circumference of the glass — a mark between 5 mm and 10 mm will be quite sufficient. Place the knife on the bench and moisten the index finger of the right hand, use this

34 *Laboratory Scientific Glassblowing: A Practical Training Method*

Fig. 5.1.

Fig. 5.2.

to moisten the "mark". With the scribe mark pointing away from you, place the thumbs behind the mark while gripping the tube (Fig. 5.5) and bend the tube outwards.

The glass will break at this point (Fig. 5.6), place the cut tube on the bench and repeat the process at each mark, resulting in five lengths of glass tubing.

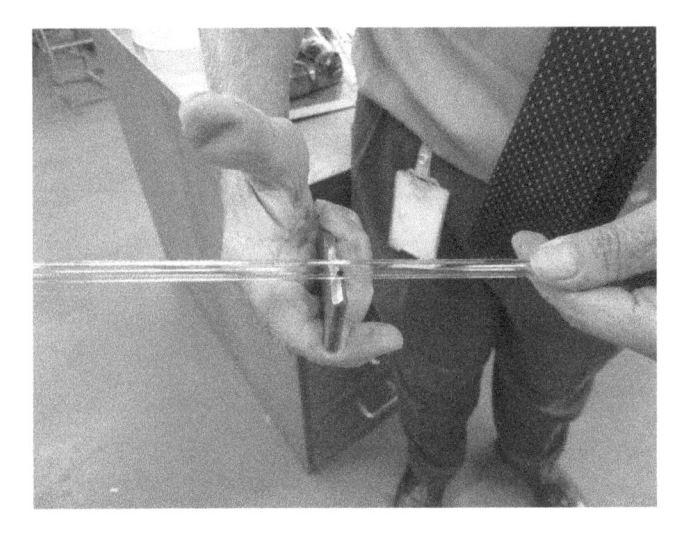

Fig. 5.3.

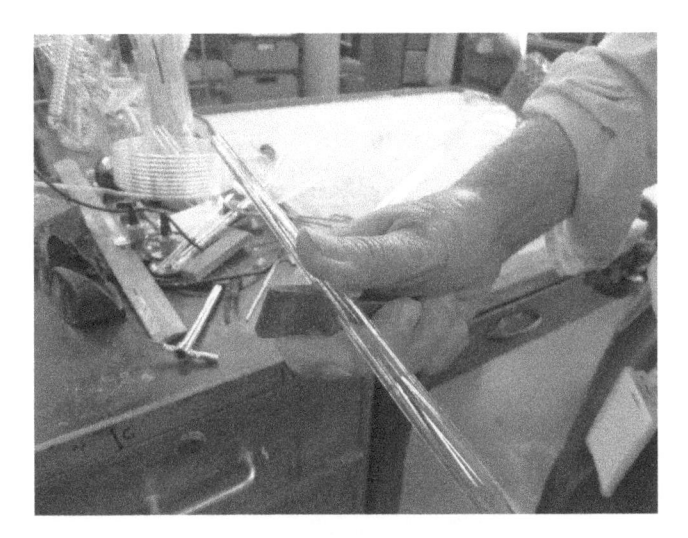

Fig. 5.4.

Light the bench burner and adjust the gas and oxygen to give a flame 12–15 mm (½″ plus) in diameter (Fig. 5.7).

Select a length of the cut tube and while rotating position the end of the tube at the edge of the flame holding the glass at right angle to the flame (Fig. 5.8), a bright yellow flame will appear from the glass — which is sodium

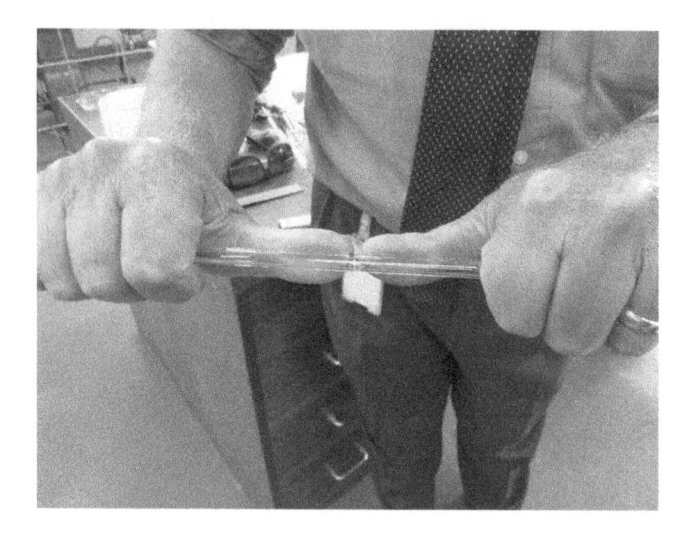

Fig. 5.5.

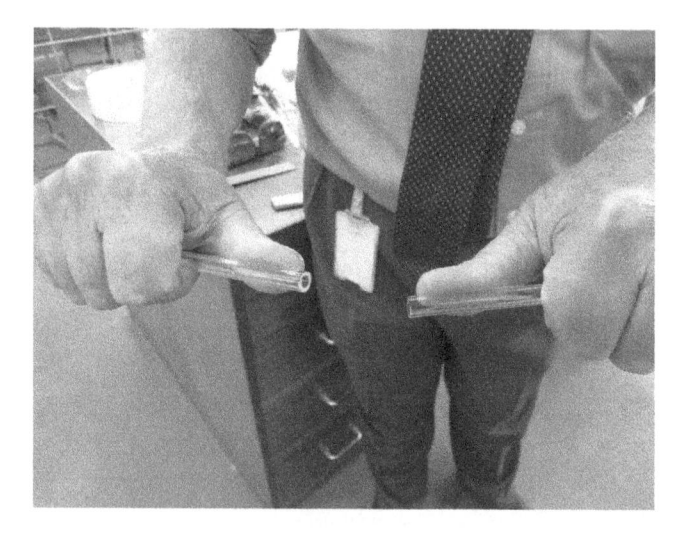

Fig. 5.6.

being burnt from the glass which indicates that the glass is molten at the point of contact.

One may be able to see that the very edge of the tube glows — withdraw the glass from the flame and allow to cool while resting the glass on a wooden support (Fig. 5.9) — and the glass is allowed to cool down naturally without

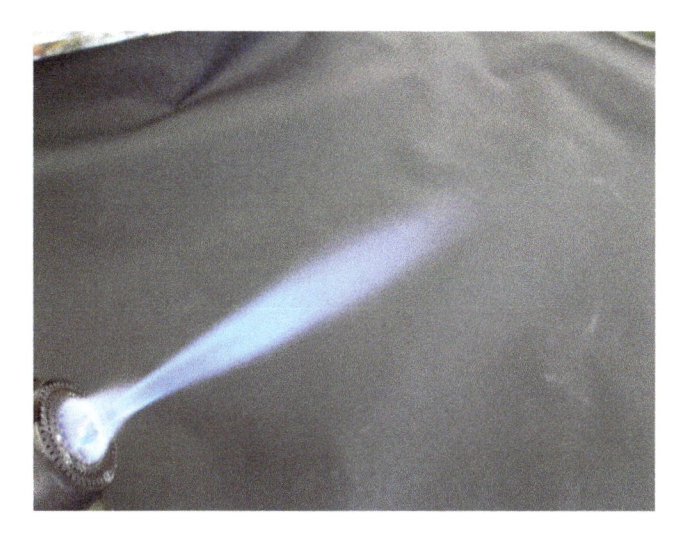

Fig. 5.7.

Fig. 5.8.

being in contact with any other material. Once cool, repeat the operation at the opposite end. Remembering that the open end should be right angled to the flame, do not angle the glass into the flame as hot gasses may well traverse the tube.

38 *Laboratory Scientific Glassblowing: A Practical Training Method*

Fig. 5.9.

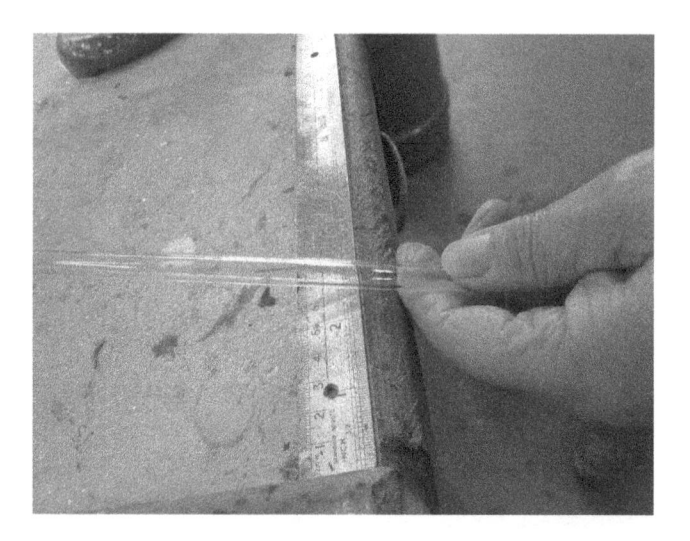

Fig. 5.10.

We have now produced five lengths of tube cut to a specific size with the ends "flame polished".

Select one of the tubes and mark the centre point with a fibre pen.

Hold the tubing in the left hand with the thumb uppermost, the tip of the thumb resting in line with the edge of the bench, the first finger will be underneath against the edge of the bench (Fig. 5.10) and the glass is held at

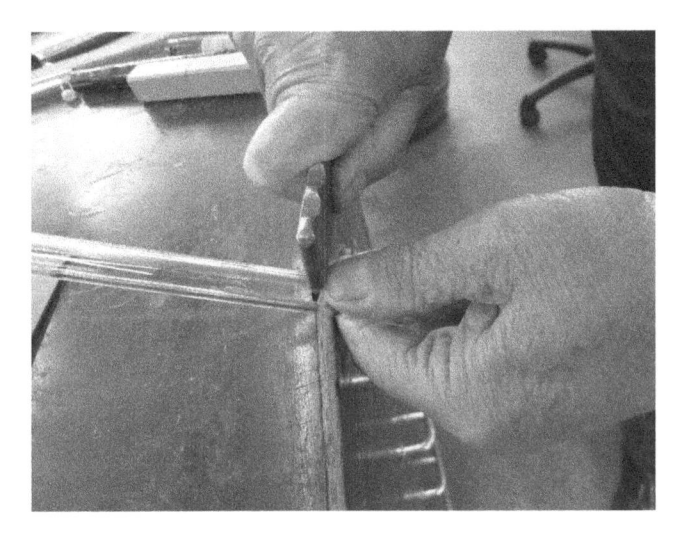

Fig. 5.11.

a slight angle, rather than lying flat on the bench, hold the glass knife in the right hand with the blade facing downwards.

Rest the blade on top of the glass while up against the tip of the thumb (Fig. 5.11).

Rotate the glass towards you while pushing the blade in the opposite direction and exerting slight downward pressure. This method will again produce a scribe mark upon the glass (Fig. 5.12). Proceed as previously described to moisten the mark and with the thumbs behind the mark bend the glass outwards to snap cut the tube, flame polish to complete.

The "thumbs "method is suitable for snap cutting smaller diameters of tubing. Larger size tube will require a variety of different methods. Take a piece of 15 mm diameter tubing, 1500 mm standard length, with the knife scribe a mark at the midpoint. We are dealing with a long length of tubing, therefore, stand up and grip the glass either side of the mark which should be facing upwards, with the arms parallel to the tubing (Fig. 5.13) "breathe out" and pull. The arms will drop and the tubing will snap, it will be noticed that greater force is required as the diameter increases, mark the mid-point on the two pieces and "snap cut" as per this method and "flame polish" the ends.

We will now use heat to cause the initial crack. Take a piece of the pre-cut 15 mm ø tube and again mark and using the knife scribe at the midpoint. Take a piece of 6 mm ø rod some 100 mm 4" long, heat the very tip of the

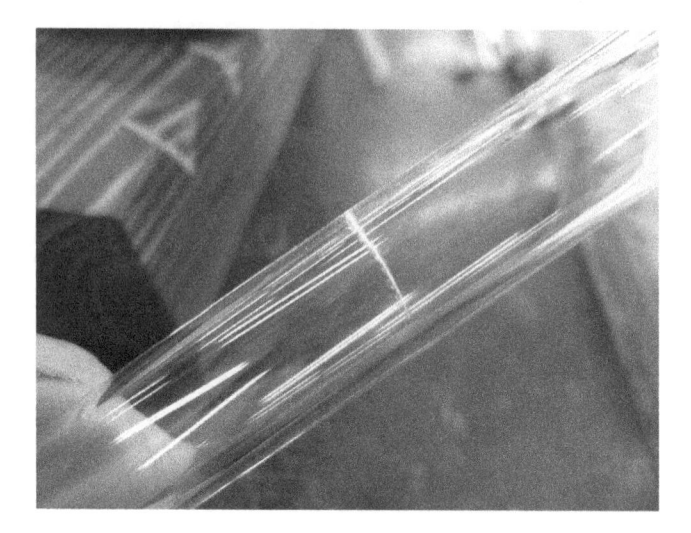

40 *Laboratory Scientific Glassblowing: A Practical Training Method*

Fig. 5.12.

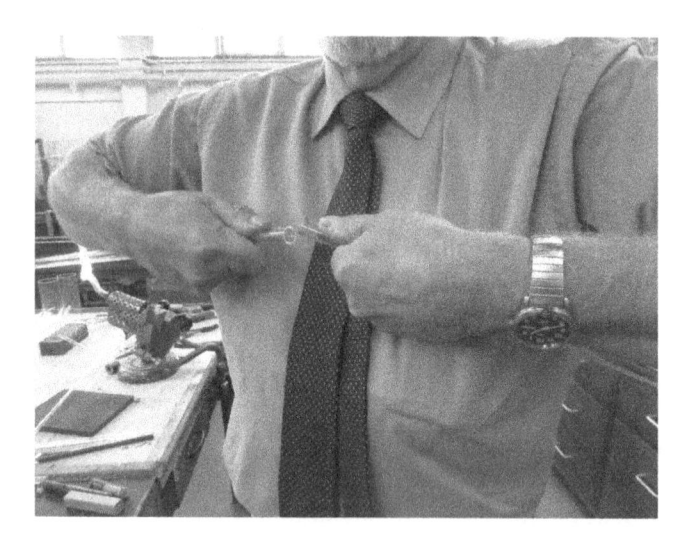

Fig. 5.13.

rod in a gas oxygen flame until molten (Fig. 5.14), remove from the flame and gently place the very tip onto the centre (Fig. 5.15) of the mark which will generate a crack sometimes all the way around the circumference of the glass, having started the crack, hold the glass using either grip described

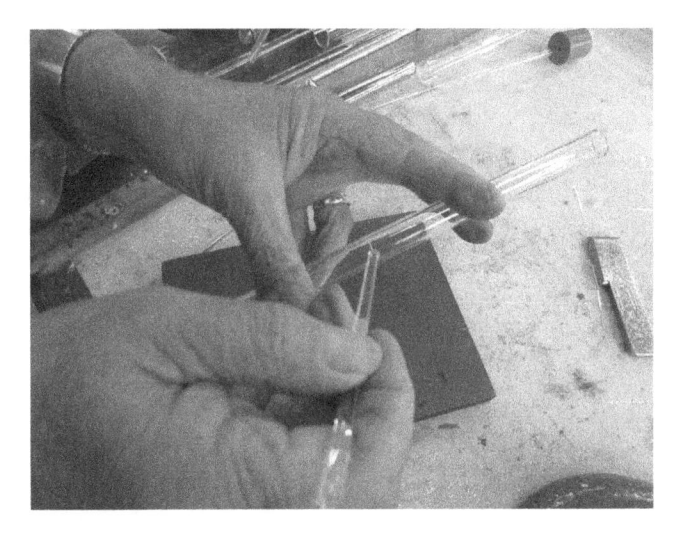

Fig. 5.14.

Fig. 5.15.

previously to encourage the crack to travel and "snap" the tube, flame polish the sharp ends.

When cutting tubing of smaller diameter which are too small a length to hold we can use yet another method.

42   *Laboratory Scientific Glassblowing: A Practical Training Method*

One will require a V block or sharp edge to complete this particular exercise. Take a piece of 8 mm or less tubing, scribe a mark 12 mm ½″ from one end (Fig. 5.16).

Hold the tubing in the left hand with the mark to the right facing upwards directly over the edge of the V block (Fig. 5.17) — moisten the

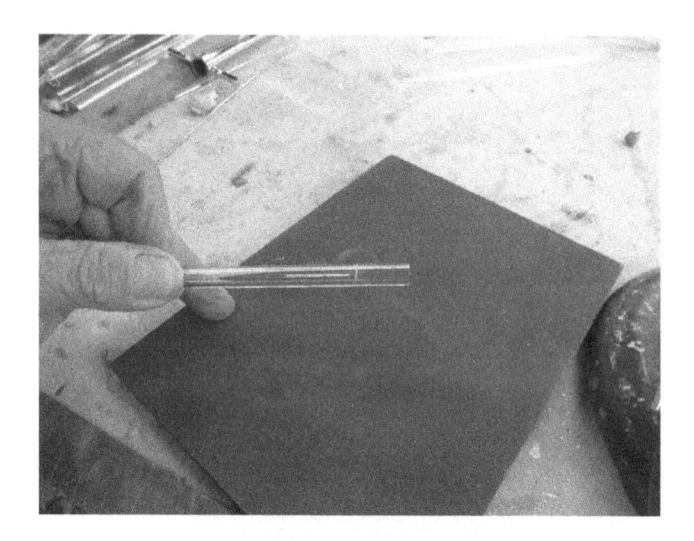

Fig. 5.16.

Fig. 5.17.

mark, hold the glass cutter in the right hand and using the back of the knife (not the blade) smartly tap the tubing to the right of the mark (Fig. 5.18). This method is ideal for opening the ends of "spear points" (Fig. 5.19).

Fig. 5.18.

Fig. 5.19.

Cutting flat glass will require an altogether different technique. The flat glass cutter has a small tungsten carbide wheel attached to a handle (Fig. 5.20).

The table on which the glass is to be cut should be "flat" ideally with a baize or non-slippery surface. The straight edge along which the cutter is guided should have a non-slip backing (Fig. 5.21) as this will be in contact with the smooth surface of the glass.

Fig. 5.20.

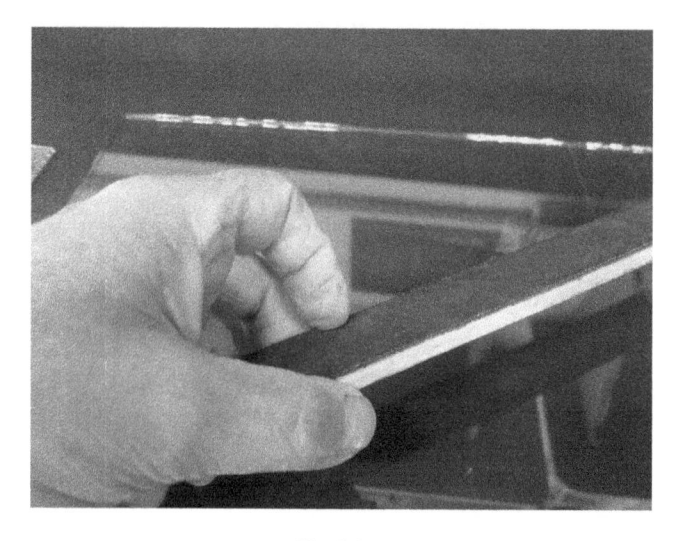

Fig. 5.21.

Hand Techniques 45

The "grip" of the wheeled cutter is important as this affects the angle at which the tungsten carbide wheel marks the glass.

The handle is placed between the first and second fingers (Fig. 5.22) while being braced with the thumb.

As one squeezes the handle the whole wrist will "lock". The handle/cutter should be held vertically (Fig. 5.23), with the glass laid on the table and

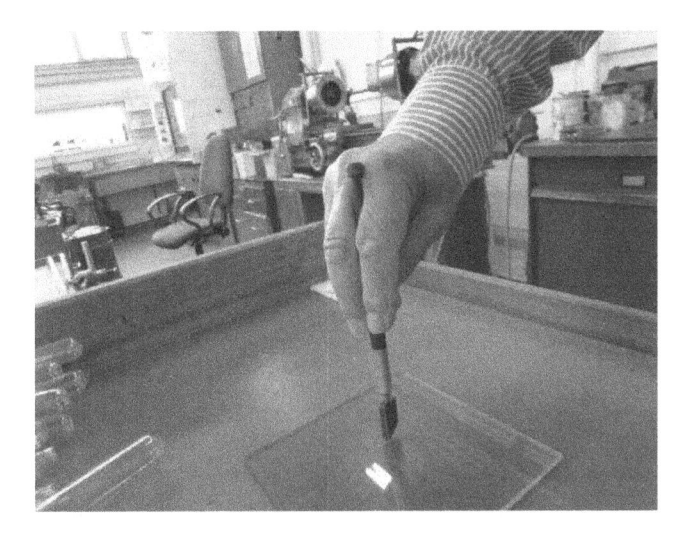

Fig. 5.22.

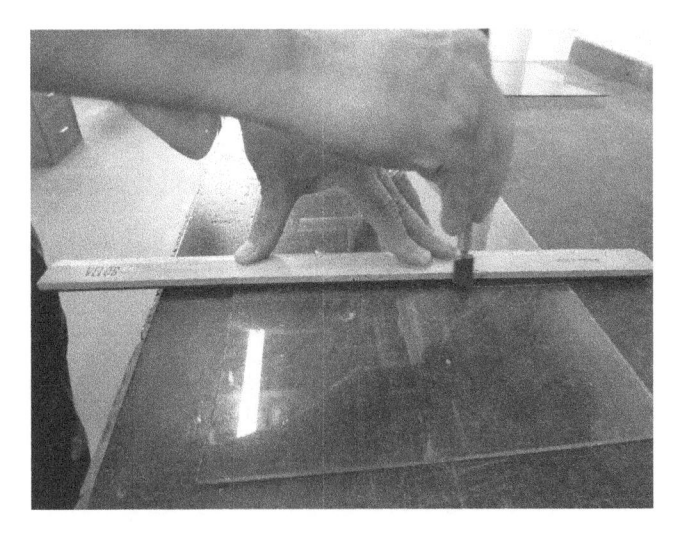

Fig. 5.23.

with the left hand resting on the glass and while standing not too close to the table, touch the far edge of glass, with the wheel draw the cutter towards the body using a constant pressure in one motion to the near edge. Carefully raise the glass and place the tip of the handle directly underneath the scribed line (Fig. 5.24). Using both hands press on either side of the mark. The glass will break along the scribed mark (Fig. 5.25). The method described will safely cut flat glass up to 6 mm thick.

Fig. 5.24.

Fig. 5.25.

Hand Techniques   47

When using a straight edge to cut glass to a specific size, firstly mark from one edge a number of times.

Place the straight edge to the left of the marks and use the glass cutter as a gauge to allow the "off set" between the wheel and its holder (Fig. 5.26). Only when confident that the wheel is in line with the measured marks press hard down on the straight edge and draw the cutter towards the body. When longer lengths are to be cut the straight edge can be secured using suction cups (Fig. 5.27).

Fig. 5.26.

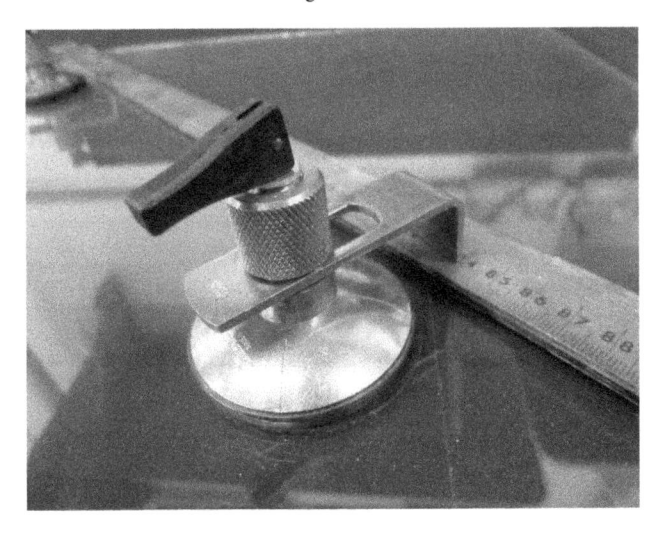

Fig. 5.27.

Fig. 5.28.

Wired glass can be cut using the above method — after marking one side only and placing the end of the handle directly under the scribed mark and the glass cracked by pressure on either side of the line. The glass will still be intact held together with the wire, with the excess piece overhanging the table edge grip the excess piece and gently raise and lower the glass which will break the wire within (Fig. 5.28).

It is always advisable to use leather palmed gloves to complete the above action.

When cutting laminated glass scribe <u>both</u> sides, break the glass using the end of the glass cutter handle as previously described. The laminate will hold the glass in place. Slice through the plastic laminate with a sharp knife to release each piece of glass.

DO NOT UNDER ANY CIRCUMSTANCES ATTEMPT TO CUT TOUGHENED GLASS.

Toughened glass is made by introducing a thermal shock to the surface of the glass which causes a vast amount of stress over the entire surface (Fig. 5.29). Should the surface be broken or scratched the glass will stress relieve itself by disintegrating suddenly.

There may be identification marks for toughened glass — but these vary across the globe. One must consult one's own countries' standards to be sure.

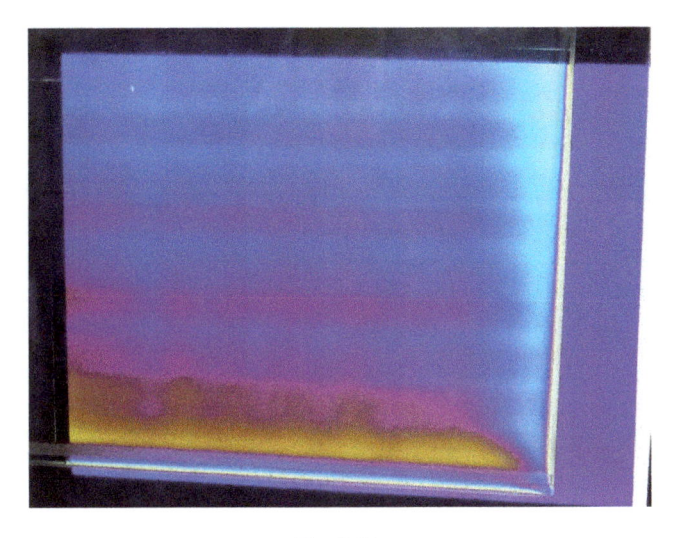

Fig. 5.29.

There may well be small indent marks along one edge where the glass had been gripped while undergoing the toughening process.

When viewed under polarised light the strain patterns are obvious.

5.2 Spear Points

There are many starting points for manipulating molten glass. It is my opinion that the best example to begin with is the drawn "Spear Point".

To begin:

Select a piece of 12 mm (millimetre) outside diameter (ø) medium-walled borosilicate tubing: a nominal half inch (½″)(ø), cut a 30 cm (centimetre) or twelve inch (12″) in length, "flame polish" each end and allow to cool.

During the "spear point" exercise it will be necessary to mimic the left and right hand positions as described in the text while referring to the photographs.

The left hand is the driving force rotating the glass constantly to overcome the effects of gravity on the molten glass.

Rest the elbow on the bench and the wrist at a right angle, support the glass tubing with the second, third and fourth fingers which are balanced by the edge of the palm (Fig. 5.30).

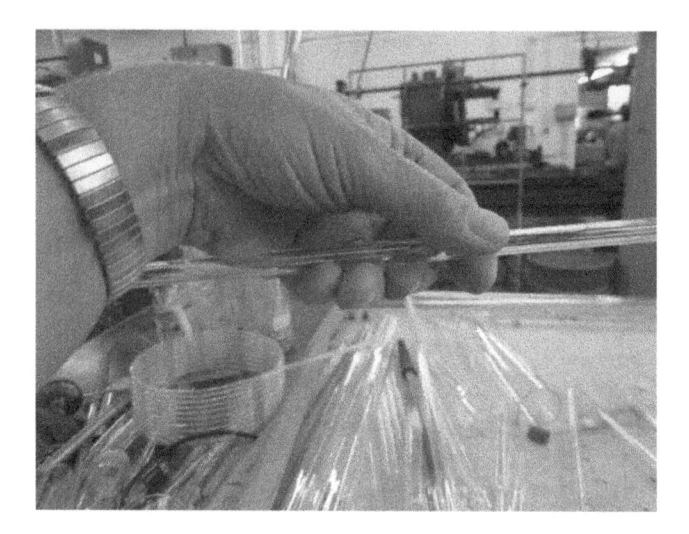

50 *Laboratory Scientific Glassblowing: A Practical Training Method*

Fig. 5.30.

Fig. 5.31.

Only the first finger and thumb rotate the glass away from the body (Fig. 5.31).

The right hand holds the glass as one would hold a pen (Fig. 5.32), with the first and second fingers supported by the thumb. Position the right hand with the palm uppermost (Fig. 5.33). The first and second fingers now act as

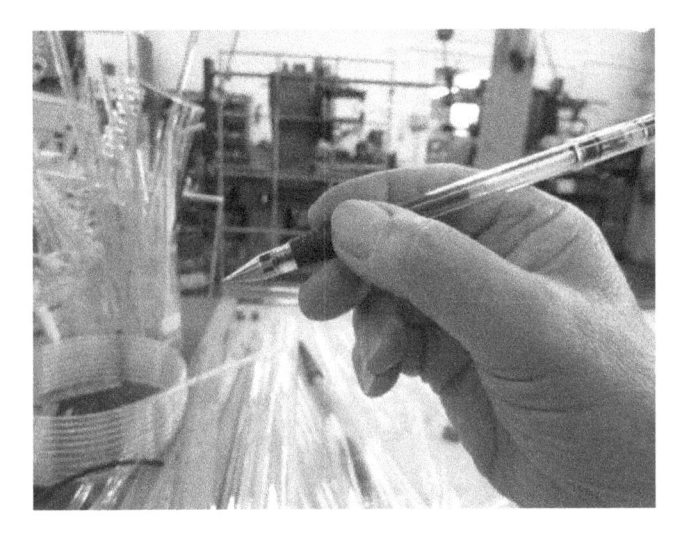

Fig. 5.32.

Fig. 5.33.

a V block, supporting the tubing while only the thumb rotates the tubing (Fig. 5.34).

Practice without the flame, rotating the glass — if viewed from the right end one will be rotating "clockwise" (Fig. 5.35).

Fig. 5.34.

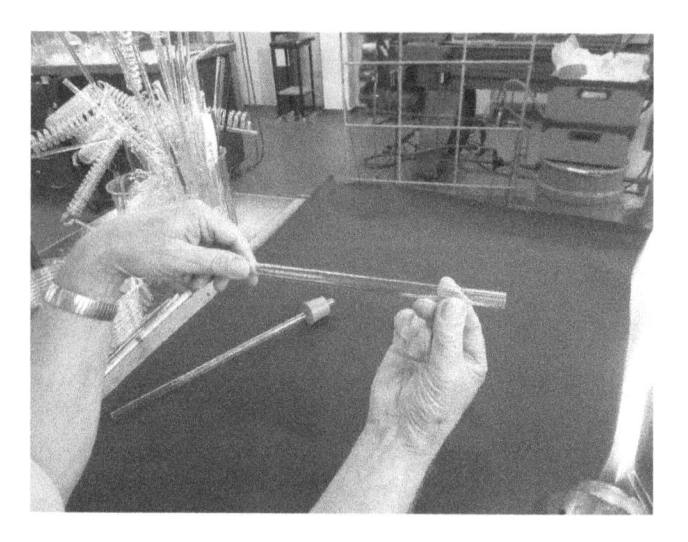

Fig. 5.35.

Adjust the burner to give a flame 25 mm wide (1″), using gas/oxygen and air if available for borosilicate glass.

Gas and air for soda glass — check that the hand positions are correct with the left hand "over the top" and the right hand underneath (Fig. 5.36).

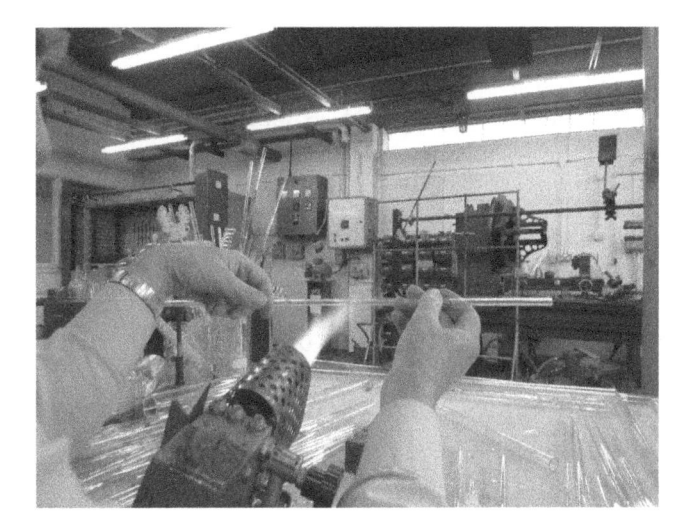

Fig. 5.36.

Fig. 5.37.

Rotate the glass as it enters the flame at a point two thirds its length from the burner tip — work in the top of the flame where the gasses are fully combusted which in turn will heat the glass evenly (Fig. 5.37).

Those using soda glass will need to have the glass enter from the top of the flame after first warming the glass in the hot air zone above the flame.

Soda glass is less able to withstand too sudden a thermal shock. Within seconds the glass will start to soften. Continue rotating the glass as this is the only way to control the softening glass. As soon as the glass softens there will be a natural tendency to withdraw the glass from the flame as it changes from solid to liquid. Rather than take the glass from the flame, raise the glass to the cooler tip of the flame.

Concentrate on using the thumbs to rotate the glass. The student may find it helpful to tap the floor with their foot in time with the thumbs. Keep the glass rotating in the flame for a count of 10 which is probably longer than one would wish. Raise the tube just out of the flame whilst still rotating (Fig. 5.38), allow the glass to cool by counting to five. One will feel the glass starting to stiffen at this point, pull the glass apart rotating all the time which should produce a slender section of tube 2 mm in diameter (Fig. 5.39). Continue to rotate until the glass is completely stiff. Touch the centre of the drawn tube with the edge of the flame while pulling apart (Fig. 5.40). This will cause the glass to melt and separate to form two "spear points". Allow to cool before resting the tip of the spear point against the edge of a V block. Scribe a mark with a glass knife, break off the tip by tapping smartly with the knife (Fig. 5.41). Leaving a hole through which we will blow and through which hot gasses will escape during the following process.

Fig. 5.38.

Fig. 5.39.

Fig. 5.40.

Holding the 12 mm ø tube in the left hand and the "spear point" in the right return the glass to the flame by aiming 25 mm (1″) to the left of the "shoulder". Rotate the glass in the flame again for a count of 10, withdraw from the flame while still rotating count to five and then pull apart (Fig. 5.42), allow to cool.

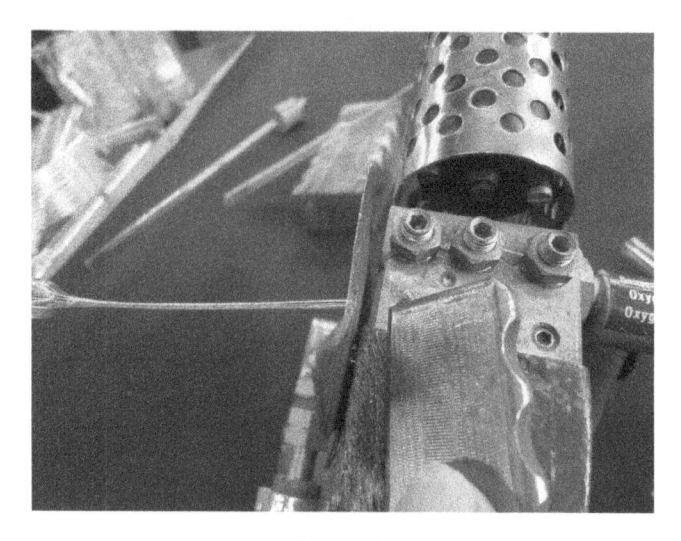

56 *Laboratory Scientific Glassblowing: A Practical Training Method*

Fig. 5.41.

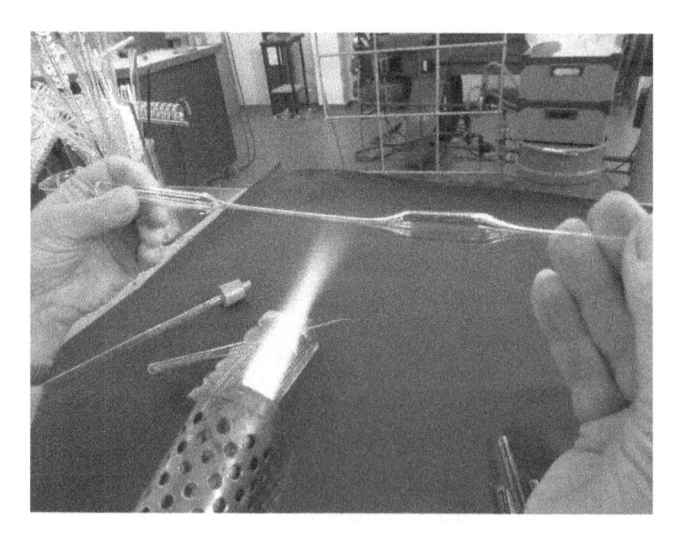

Fig. 5.42.

Place the open ended spear point in the left hand, reduce the width of the flame to 12 mm (½″), aim the flame at the point where the "spear point" meets the tubing ("The Shoulder"), while rotating raise the left hand and continue to rotate while the glass is at an angle of 30° (Fig. 5.43).

As the glass softens on the "shoulder" pull the right hand away which will allow the flame to burn through — remove the "spear point"(Fig. 5.44).

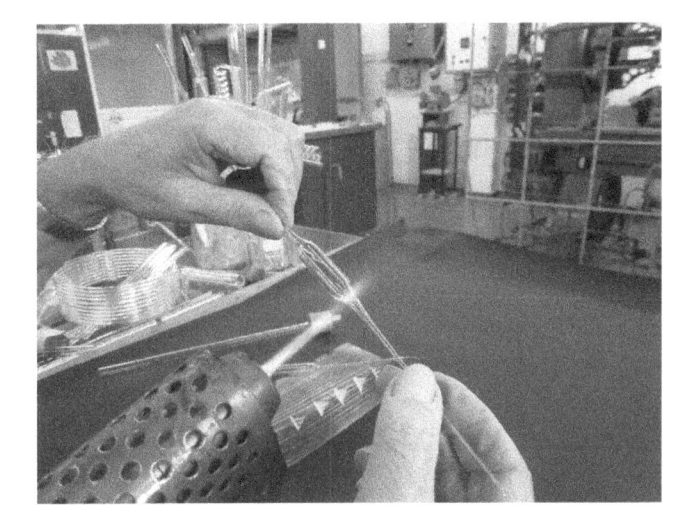

Fig. 5.43.

Fig. 5.44.

Continue to rotate the glass in the left hand and raise the left hand until the glass is at 45° (Fig. 5.45). Aim the flame at the end of the now sealed tube which should now be quite molten. Withdraw the glass from flame, rotate for a count of three, blow down the open ended spear point. The aim is to make a hemispherical end known as a "test tube" or "domed end" (Fig. 5.46).

58 *Laboratory Scientific Glassblowing: A Practical Training Method*

Fig. 5.45.

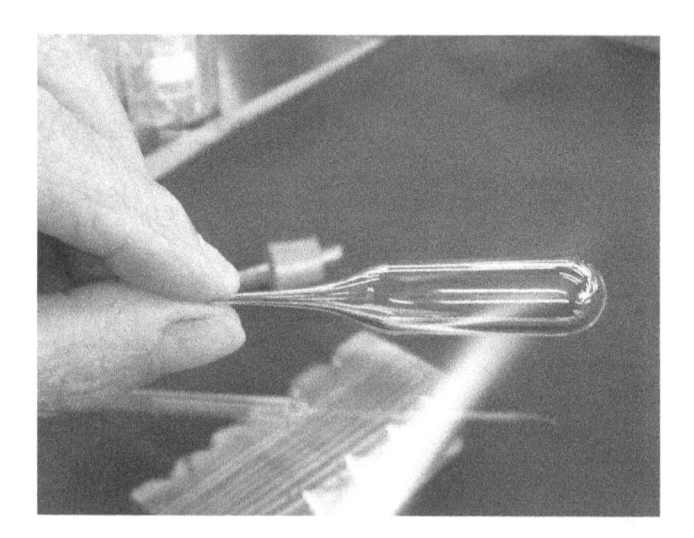

Fig. 5.46.

The hotter the glass is the less pressure is needed to form the "domed end". With practice, blowing as the glass is cooling will give a good shape and an even wall thickness. The "domed" end may be returned to the flame, reheated and blown again until satisfied with the result.

The importance of being able to pull a strong straight "spear point" cannot be overemphasised. The "control" of a molten material is true hand skill, once mastered gives an understanding of how glass feels and reacts. The exercise also gets the students' hands used to the radiated heat from both the glass and the flame. One also becomes more proficient at the controls of the burner.

5.3 Straight Joins

The following two exercises will be of benefit if practised prior to the actual "straight join" and can be used at any time to practise the left and right hand actions.

Select a rubber bung of 1 inch 25 mm diameter. Cut and flame polish a 20 cm length of 9 mm outside diameter medium walled tubing.

Select a cork borer through which the 9 mm will slide. Mark the centre of the rubber bung at either end using the tool described (Fig. 5.47) made from either metal or wood where a 90° angle is dissected. This simple device when placed against a circle will find the centre when scribed and after positioning at various points around each end of the bung. One is now confident of the centre (Fig. 5.48).

Fig. 5.47.

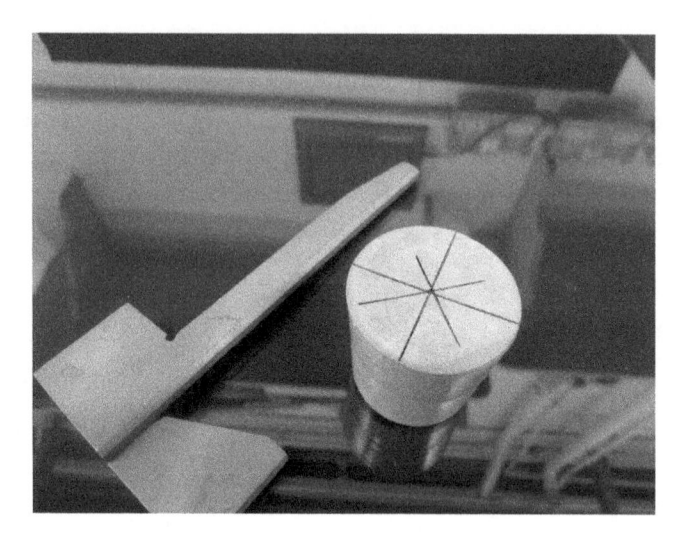

Fig. 5.48.

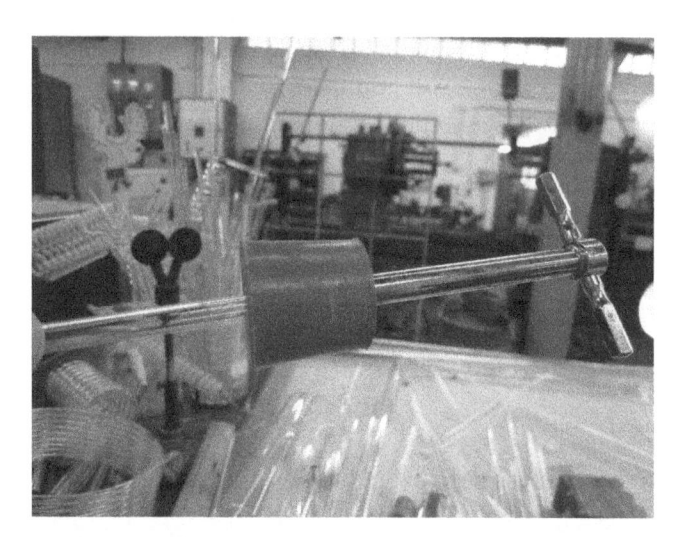

Fig. 5.49.

Using the cork borer well lubricated with water from each end of the bung remove the core. Leave the cork borer in place, insert the 9 mm ø tube through the borer (Fig. 5.49). Remove the borer leaving the 9 mm ø glass tube in place which is held snuggly by the rubber bung.

Fig. 5.50.

The above method avoids the need to "force" a glass tube through the bung.

The exercise proper: Hold the glass in the left hand, practise rotating using only the thumb and the forefinger, while supporting the weight with the second, third and fourth fingers combined with the edge of the hand. The elbow is rested upon the bench (Fig. 5.50).

The exercise will develop control, strength and a "feel" because the weight of the bung will have its own momentum. Note that the tube and bung will be used later to hold larger tubing.

The second exercise: Select a piece of 12 mm ø tubing with a bore of 9 mm cut to a length of 25 cm — select a 25 cm length of 8 mm ø tubing. Using a fibre pen mark a line along the length of each tube to a distance of 4 cm, mark the circumference (Fig. 5.51). The larger diameter tube is held in the left hand, rotate using the thumb and index finger supported with the second, third and fourth fingers. Insert the smaller 8 mm ø tube inside the larger to the marked 4 cm depth — align both horizontal lines (Fig. 5.52).

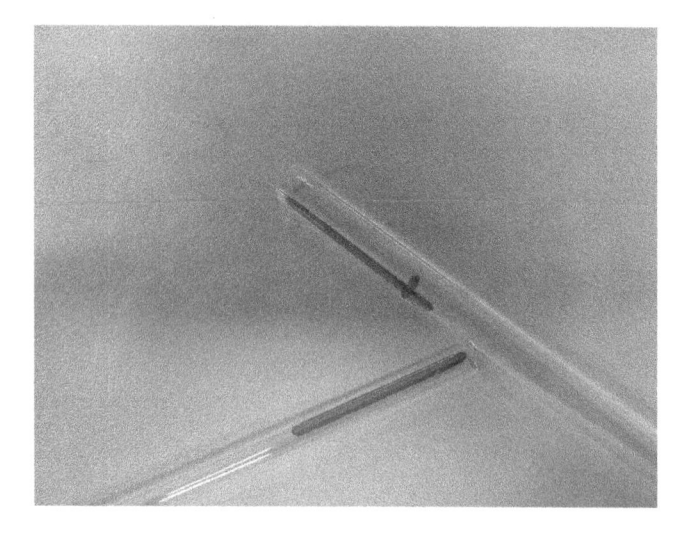

62 *Laboratory Scientific Glassblowing: A Practical Training Method*

Fig. 5.51.

Fig. 5.52.

Rotate the tubes together keeping the marks aligned to the same depth. This exercise will simulate what will happen when making a straight join: the solid tubes in either hand with a liquid in the centre. A great deal of concentration will be needed to keep both lines together.

The exercise enables one to practise rotating in unison without the fear of losing control of the glass.

The straight join: Select a piece of tubing 10 mm ø. Cut two pieces 160 mm long and flame polish the ends and allow to cool. Seal one end with a cork and place in the left hand using the "over hand" grip we have practised earlier. Adjust the oxygenated flame to a width of 12 mm. Hold the open-ended tube in the right hand from underneath as we have practised earlier. Bring the edges of the tubing to the very edge of the flame (Fig. 5.53). Rotate the glass or rotate back and forth if this is found to be easier. Watch the orange sodium flame coming off the very edges of each tube which indicates that the glass is melting evenly. Withdraw the glass towards you out of the flame by 20 mm (Fig. 5.54). Touch the edges together, give a light pull apart and blow.

Touch, Pull, Blow

The join is virtually made. There will probably be a slight ridge at the join after this "initial join". Concentrate on "rotating" and return the glass to the flame. Soon the glass will start to soften, keep the hands together while rotating, do not allow them to pull or wander apart.

The left hand must be the dominant hand rotating constantly and evenly. The right hand has a much lighter grip. Once the glass can be seen and "felt" to be molten withdraw from the flame, once more rotate to a count

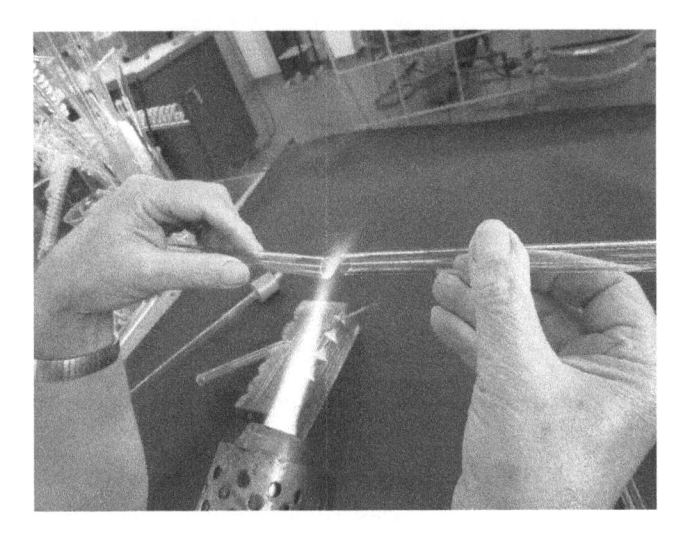

Fig. 5.53.

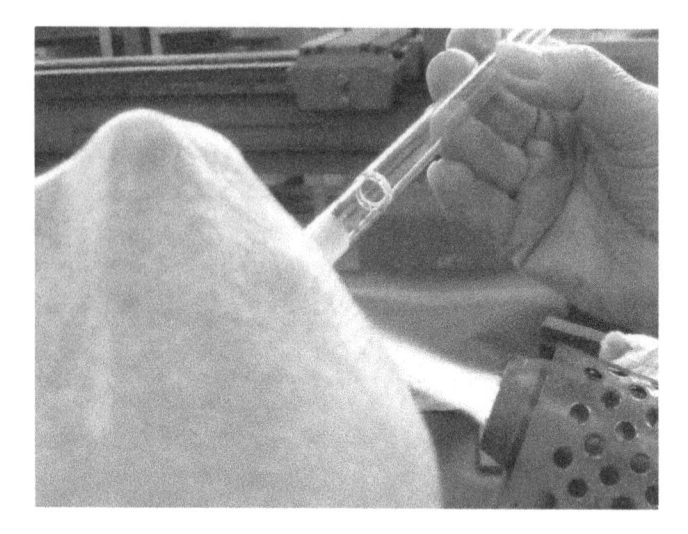

Fig. 5.54.

Fig. 5.55.

of four bring the open tube to the mouth and blow making the seal with lips only (Fig. 5.55).

The distance between each hand must be constant when either melting or blowing. Watch carefully the glass while blowing — control the blow to retain the tube diameter.

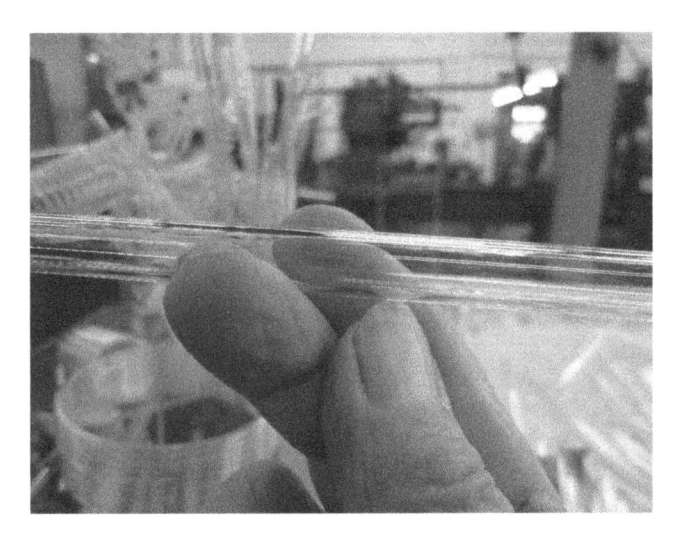

Fig. 5.56.

Return to the flame — rotate and melt — withdraw and pause.

Then blow as the tube cools to create a slight bulge or "over blow". As the glass cools very gently pull apart whilst rotating to regain the original diameter.

The exercise is complete, put this first attempt to one side while we practise further. We will compare this later after gaining a more complete understanding of the technique. The process of having the left hand dominant and the right hand in support will centralise the softened/molten/cooling glass. With practice one will be able to feel the glass centralise itself. If as the glass cools we rotate only with the left hand and hold the tube to the right with the first and second fingers acting as a "V block" (Fig. 5.56).

The most important part to note is "wall thickness". Look "into" the glass for the grey line parallel to the tube. First look at the end of the tube and see the wall thickness, then hold the tube sideways against both a light and dark background. It is the wall thickness that is so important for you to be able to recognise (Fig. 5.57), it will indicate uneven thickening and thinning of the glass wall. In use an even wall thickness is what is to be aimed for, a reduction in wall thickness reduces the strength and the ability to withstand internal pressure.

Do not be bemused or overawed by the shiny surface of glass.

Look through the glass.

See Safe Working Pressure Chart in Chapter 4.

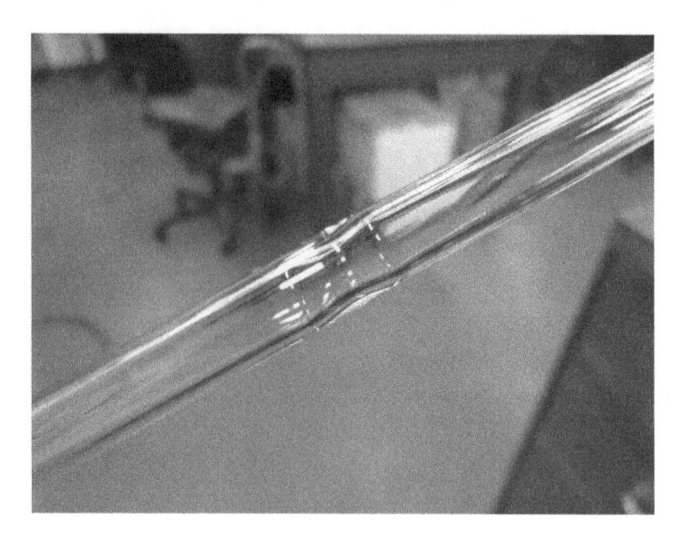

Fig. 5.57.

5.4 Unequal Joins

Rounded Shoulder

This is in essence a straight join once the larger diameter is reduced to the smaller diameter.

Cut a 400 mm long piece of 15 mm ø tubing and flame polish the ends and allow to cool, cut two pieces of 8 mm ø tubing 150 mm in length, flame polish the ends and allow to cool.

Select a cork or rubber bung that will fit and seal the 15 mm ø tube.

Mark to the midpoint with a fibre pen of the larger 15 mm ø tube.

Hold the glass in the manner as described with the left hand thumb and first finger to rotate as the second, third and fourth fingers support. The right hand underneath first and second fingers supporting while the thumb rotates the glass (Fig. 5.58).

Select oxygen, gas air flame 30 mm wide. Rotate the glass and enter the flame near the top at the marked position.

We will be pulling a spear point, the aim is to produce a point that lies along the central axis with an even wall thickness which is robust and 300 mm in length.

Touch the drawn tube on the edge of the flame at the midpoint and pull apart to produce two spear points, allow to cool (Fig. 5.59).

Fig. 5.58.

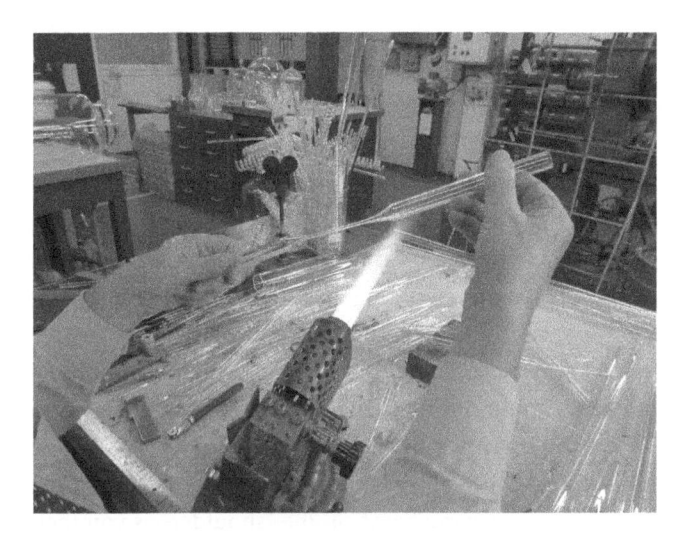

Fig. 5.59.

When cool inspect the point by rotating the 15 mm ø tube in the left hand while holding horizontally, look carefully at the area where the point is drawn from the tube ("The Shoulder") which should be even and central.

Rotate with the left hand while resting the spear between the first and second fingers of the right hand (Fig. 5.60).

68   *Laboratory Scientific Glassblowing: A Practical Training Method*

Fig. 5.60.

One will see and "feel" the spear is running true and central, if so, congratulations.

If not running centrally adjust the flame to the same diameter as the tube (15 mm).

Rotate the glass in the left hand, run the spear point between the first and second fingers of the right hand and introduce the glass to the very tip of the flame which is directed on the "shoulder" only and rotated evenly.

Do not use the thumb of the right hand but "feel" the spear point centralise — continue to rotate and withdraw from the flame, as the glass cools and stiffens start to use the thumb of the right hand as it solidifies. The spear point should be far more central than at first. It is important not to overheat the shoulder during this operation. Allow to cool.

Select the second piece of 15 ø tubing, hold the tubing in the left hand and rotate the glass — play the flame on the "shoulder" as you incline the left hand to an angle of 45° which will mean that the bottom edge of the shoulder will lie horizontally (Fig. 5.61).

As the shoulder softens draw the right hand away while keeping the left hand in position. The spear point will thin and burn through leaving only the tube and shoulder (Fig. 5.62).

Remove the glass from the flame and return to the horizontal. Readjust the flame to less than 10 mm diameter. Introduce the tip of the shoulder to

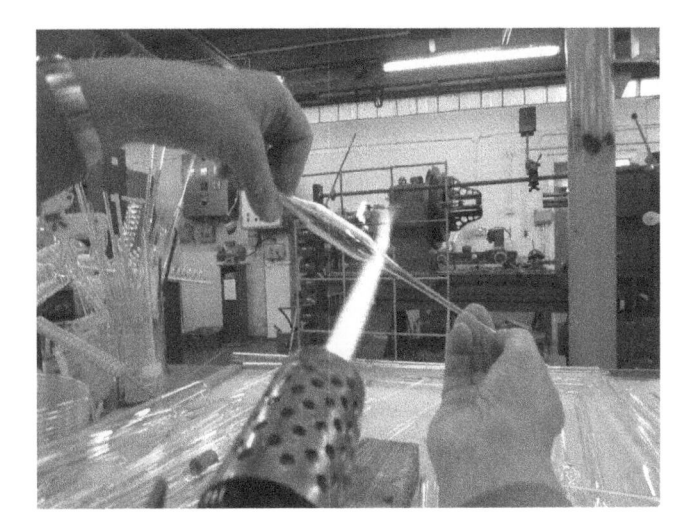

Fig. 5.61.

Fig. 5.62.

the edge of the flame — watch the tip melt, withdraw from the flame and blow to produce a thin bubble (Fig. 5.63) — use the glass knife to remove the bubble at the shoulder with a slicing action to reveal a hole (Fig. 5.64), which we would aim to be something similar to the diameter of the tube we are about to join. Seal the end of the 15 mm ø tube with the cork or rubber

70 *Laboratory Scientific Glassblowing: A Practical Training Method*

Fig. 5.63.

Fig. 5.64.

bung. Adjust the flame back to 15 mm wide. Hold the tube in the left hand and the smaller 8 mm ø tube in the right, bring both open ends of the tubes to either side of the flame, "rotate" evenly as the edges melt, once hot, withdraw from the flame towards you by 50 mm, touch the edges together, pull slightly and blow without a pause (Fig. 5.65).

Incline the left hand and support in line underneath with the right hand. Aim the flame on the shoulder. The right edge of the flame should be heating

Fig. 5.65.

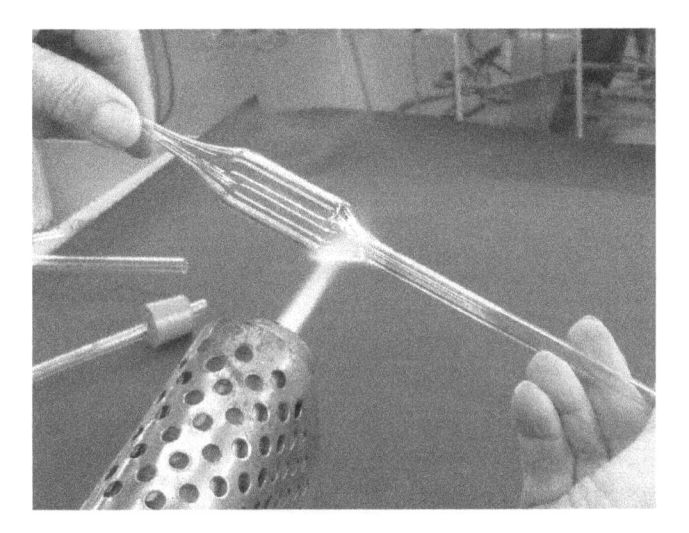

Fig. 5.66.

the initial join. As the glass melts it will be dropping under gravity, hence the new hand position (Fig. 5.66).

Watch and feel the glass melt, withdraw from the flame, pause for a count of three, then drop the left hand to the vertical while keeping the right hand inline and at an equal distance apart. Blow and watch the shoulder increase in size and form the rounded shoulder (Fig. 5.67), return to the horizontal and rotate the glass while supporting only with the first and second

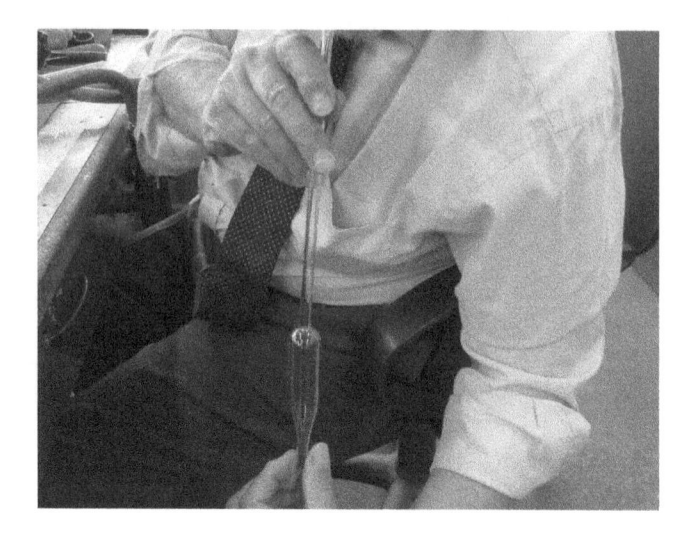

Fig. 5.67.

fingers of the right hand to confirm that both tubes are running along the same axis. If not, one may reduce the flame diameter and by heating the shoulder slightly centralise the tubes using the method previously shown to centralise the spear point.

This exercise can be repeated by drawing a spear point just to the left of the shoulder and starting again.

One may have found it difficult to rotate the larger tube in the left hand, be assured this will improve with time.

One may also use the spear point if strong and central as a handle, for example:

Select the first tube and spear point that we have centralised previously, mark the tubing 50 mm from the shoulder. Open the tip of the spear point. Hold the spear point in the left hand, once more using a flame 30 mm wide heat the tube at the mark until soft and workable, withdraw from the flame, pause and pull apart to produce the double spear 300 mm long. Separate the two points using the edge of the flame and allow to cool a little.

Reduce the diameter of the flame to 15 mm wide, incline the glass while burning off the spear point, reduce the flame heat and blow out the end of the shoulder to produce a hole in preparation for the unequal join.

Using a strong straight spear point will make it easier to rotate the glass using the smaller diameter tube between the finger and thumb of the right hand.

Once the unequal join has been completed allow to cool, hold the glass in the left hand by the 8 mm ø tube and pull off the spear point at the shoulder and proceed to make a domed or test tube end as described in Section 5.2.

Unequal Join Tapered Shoulder

Using both methods previously described proceed to the stage when we are about to join the larger and smaller tubes together. Make the "initial join" by touching together followed by a slight pull and blow. Incline the tube at an angle so that the shoulder between the 15 mm and 8 mm tubing is horizontal. Using the 15 mm wide flame, the right side of which plays upon the initial join, when the glass is workable withdraw from the flame while retaining the angle of inclination rotate to a count of eight and then blow in the horizontal position, do not overblow and do not allow the hands to pull apart at this stage. Rotate and observe the taper, only as the glass starts to stiffen would one pull the taper into shape while still rotating (Fig. 5.68).

It is very important to control the position of the glass in the flame as it is so easy for the smaller diameter tube to melt before the larger has got to the working temperature.

Repeat the exercises using both methods to gain confidence.

Fig. 5.68.

5.5 T Pieces

Select a length of 10 mm ø tube and cut two lengths of 200 mm long and two lengths of 100 mm long and fire polish the ends. Mark the midpoint of the longer length, adjust the burner to give a gas, oxygen flame 5 mm wide. Seal one end with a cork or rubber bung, place in the left hand and hold the tube horizontally, aim the tip of the flame at the mark. Hold the glass steady for a count of four, one will see the glass starting to melt, remove from the flame and blow down the open end of the tube which should produce a dome — return the glass to the flame and aim the flame at the tip of the dome which will soon flatten, withdraw from the flame, blow once more to produce a thin bubble of glass (Fig. 5.69). Using the back of the glass knife break off the bubble with a slicing action to reveal a hole (Fig. 5.70) 8, 9 or 10 mm in diameter. Seal the tube with a second bung. Adjust the flame to approximately 12 mm diameter, hold the longer tube slightly to the rear of the flame. With the hole facing you, hold the shorter length of tubing in the right hand as one would hold a pen and rotate in the edge of the flame while also allowing the flame to lap across the newly made hole (Fig. 5.71), when both edges are molten remove from the flame and touch the tube in the right hand against the hole, give a slight pull and blow to confirm that the "touch together" was accurate and

Fig. 5.69.

Fig. 5.70.

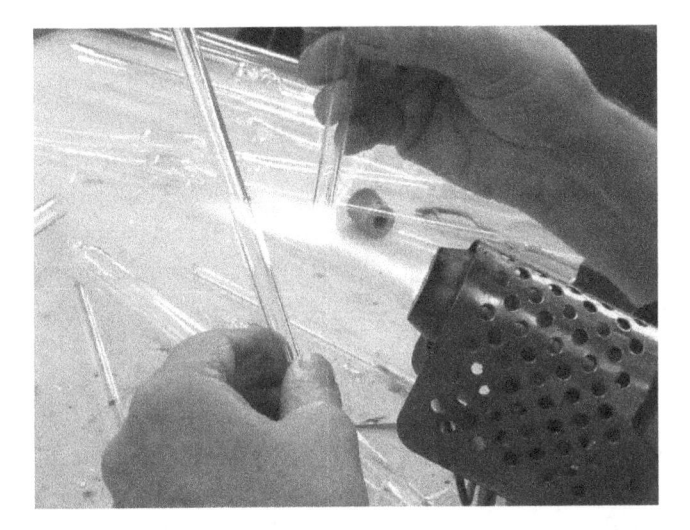

Fig. 5.71.

has made a seal between the two pieces — if successful we now have the options to complete.

Method 1

Adjust the width of the flame to 5 mm and aim the flame into the corner of the right angle while holding the glass so that as the glass melts it sags and

Fig. 5.72.

flows together to produce a radius join due to both gravity and surface tension (Fig. 5.72). Very little blowing will be required if using a hot flame although a blow to retain the diameter is recommended as one withdraws the glass from the flame to allow the glass to cool. The unmelted opposite side of the T piece will give support while working each side of the piece, each corner in turn is worked finishing with the sides. As this method is more akin to welding the amount of stain produced makes it essential to pre-warm and anneal the whole piece without delay.

Method 2

Repeat the previous stages as far as the initial join — and the "T" formed. This method is more like the "straight join" especially for the right hand which holds the smaller length of tube in the pen grip — thumb first and second finger. This time the rotation will be back and forth in line with the join as left hand rotates back and forth in large sweeping action while gripping one end of the longer tube.

The left hand must remain dominant while the right hand follows its every move as the join melts.

It is not possible to rotate in one direction so the back and forth motion must complete a full circle while mentally concentrate the flame on the far side of the join (Fig. 5.73).

Fig. 5.73.

One will be aware that the whole join is molten — remove from the flame, "pause" while holding the longer length horizontally and the shorter length vertical blow to retain the diameter repeat the melting process until all traces of the initial join have disappeared completely — this is termed "Run in".

This second method is without doubt a more difficult task but it is important to control the glass to such a degree that both methods will have to become interchangeable with the same standard of finish dependent on the size, shape and complexity of glassware design.

Both methods will reinforce one's understanding of how molten glass reacts and how to coordinate the left and right hands.

Most importantly, the second method will highlight the position of the hands in relation to the flame.

Practice both methods repeatedly until comfortable with the procedure. We will explore yet more alternative methods in future exercises.

5.6 Y Pieces

Y and T pieces have a great deal in common using similar hand techniques and flame control. Although the Y piece will require a right angle bend we will cover the basic operation here and cover the method in far more detail in the next section.

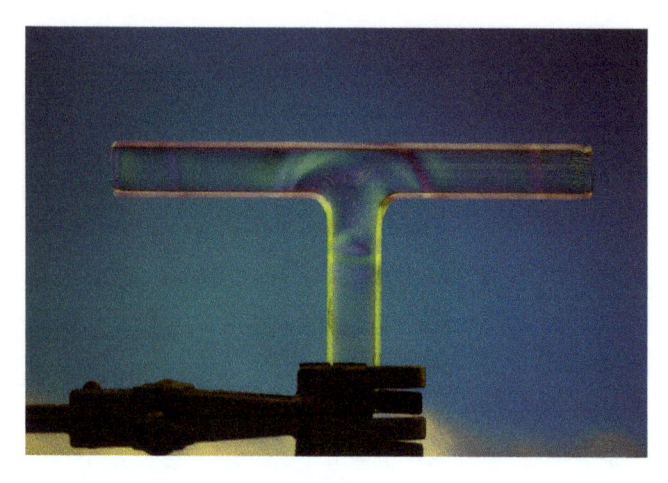

Fig. 5.74. T piece as seen under polarised light showing temperature-induced residual stain within the glass.

Select a length of 10 mm ø tubing and cut 200 mm and 100 mm lengths of tubing — flame polish each end. Take one of the longer lengths of tubing and seal one end with a cork or similar. The sealed end will be held in the left hand.

Prepare the flame which should be three times wider than the tube to be bent, in this case 30 mm wide using gas, oxygen and air to give a hot but feather-like flame (Figs. 5.74 and 5.75).

Both hands hold the glass in a similar manner. The glass rests upon the first, second and third fingers while the thumb rests on top of the tube to roll the glass back and forth along the fingers (Fig. 5.76), it is important that at least one rotation of the glass in either direction is made so as to heat and melt the glass evenly. Mark a line on the glass with a fibre pen to confirm that the back and forth rotation is taking place.

With the cork in the left hand hold the glass horizontally, rotate back and forth as the glass enters the top of the flame, continue rotating until the glass is felt to soften, count to four, withdraw from the flame raise the left hand so that the glass is in the vertical position (Fig. 5.77), hold steady with the left hand as the right hand sweeps the lower piece of glass towards you, hold the glass at right angles (Fig. 5.78) as one blows to retain the tubing diameter.

Hold this position until the glass solidifies, the bend is made so long as the hands have not drifted apart or should the glass not been sufficiently heated through the bend would be "tight" small and rippled on the inside of the bend.

Fig. 5.75.

Allow the bend to cool, readjust the flame to give a small hot 5 mm wide flame which is aimed at the outside of the bend for a count of four — withdraw from the flame and blow a slight dome, return to the flame and repeat until the dome becomes hot and flattens, withdraw and blow once more to produce a thin bubble which is sheared off with the back of the glass knife (Fig. 5.79). Use a cork or similar to seal the open tube. Readjust the flame to give a less harsh flame of 12 mm diameter. Take one of the smaller lengths of prepared tube. Hold the bend in the left hand so that the flame will brush over the edge of the hole while holding the shorter length in the opposite edge of the flame (Fig. 5.80), wait for the edges to soften and withdraw from the flame, touch the edges together, blow to confirm that the edges are sealed.

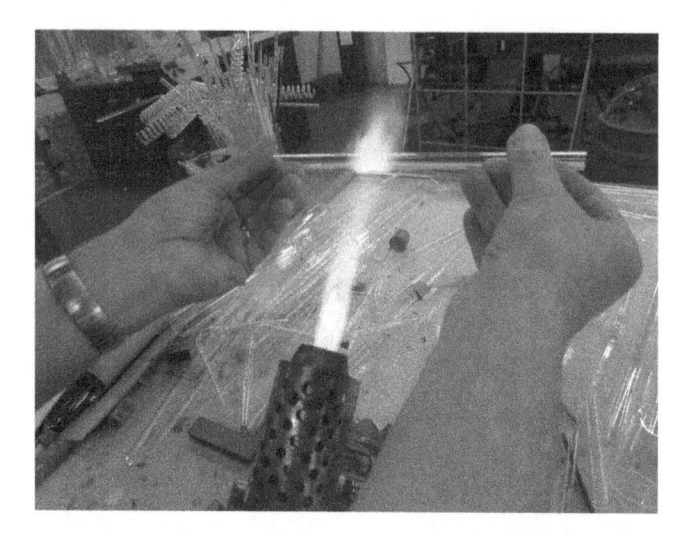

Fig. 5.76.

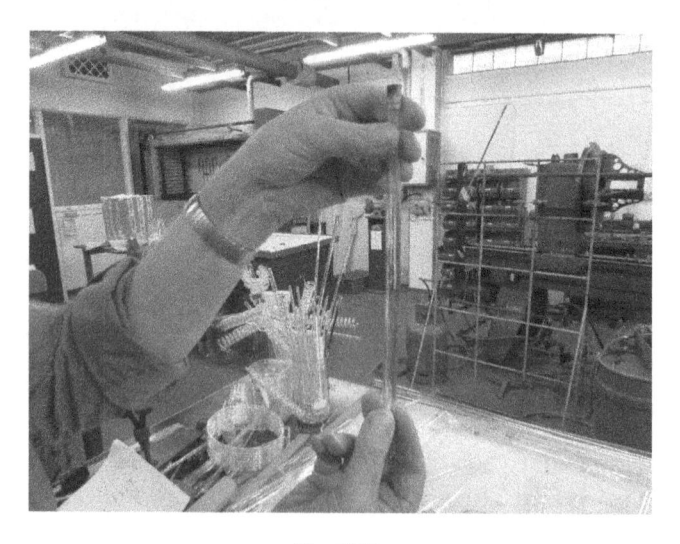

Fig. 5.77.

Reduce the flame to 5 mm diameter, proceed to "run in" or fuse the glass together using both methods as described for the T piece. One will notice quite quickly that one's fingers of the left hand are a little farther away from the flame edge. Flame control is important so as to not overmelt the glass and cause the bend to collapse and deform.

Repeat this exercise and then alternate between both T and Y pieces.

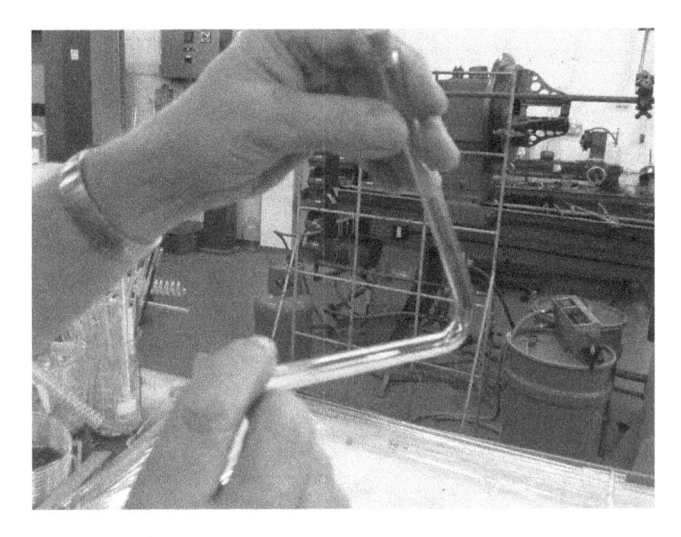

Fig. 5.78.

Fig. 5.79.

5.7 90° Bends

Select a length of the following: 6 mm, 8 mm, 10 mm and 12 mm medium walled tubing together with a 2 mm bore capillary tubing having a 9 mm outside diameter. Cut the 6 mm, 8 mm, 10 mm and 12 mm tubing into 250 mm lengths and flame polish the ends starting with the 8 mm ø tube

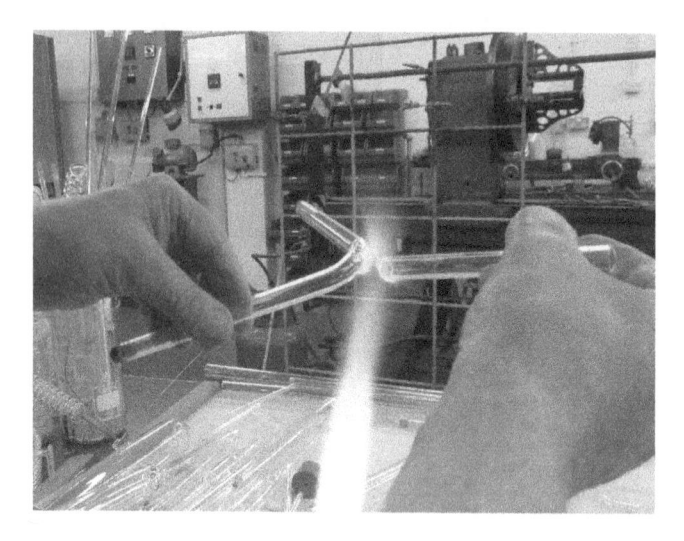

82 *Laboratory Scientific Glassblowing: A Practical Training Method*

Fig. 5.80.

Fig. 5.81.

seal one end and place in the left hand, adjust the burner to give an aerated feather-like flame. Hold the tubing underneath and rotate back and forth using the thumb and rotate the glass in the top of the flame until the glass starts to soften (Fig. 5.81), keep in the flame for a further count of five. Withdraw the glass from the flame and immediately hold the glass vertically

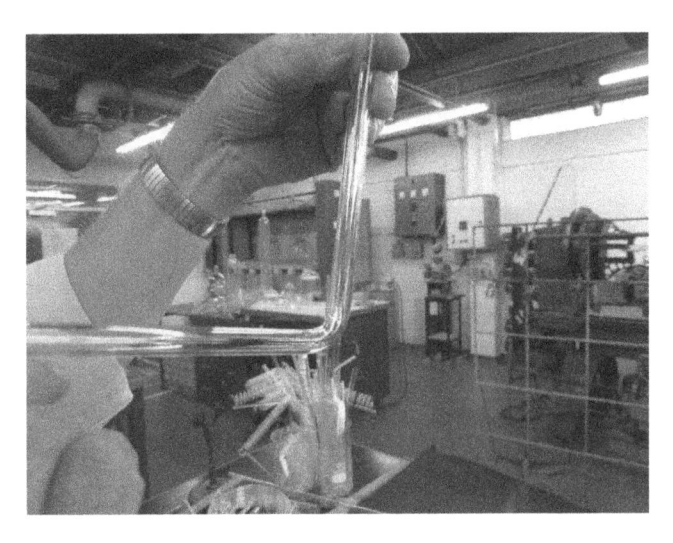

Fig. 5.82.

with the left hand, bring the right hand towards you in a sweeping motion and blow (Fig. 5.82), hold this position until the glass stiffens. Turn the glass through 90° and inspect the bend. If the bend is too sharp and has a ripple or crease in the corner of the bend, repeat the exercise with a slightly wider flame and keep the glass in the flame a little longer.

As with most glassblowing timing is everything. One must begin to feel when the glass has softened fully. One may try sliding the tubing from side to side while rotating back and forth to heat the glass evenly. Once out of the flame the left hand must hold the vertical glass rigidly while a much lighter grip with the right holds the glass. One must be sensitive as to the condition of the glass as the right hand brings the glass to the right angle. The blow must be controlled as one watches the diameter increases in size. Repeat the exercise on all 6, 8, 10 and 12 mm sizes.

Check that the bends are right angles by using a set square (Fig. 5.83).

One will be surprised just how soon it becomes automatic for you to identify and make a 90° bend. Bends can be rectified by heating the inside of the bend until softened and then gently blow to improve the shape of the bend while the outer part of the bend gives support. Bends can be resurrected but ideally if done correctly in the first place they will be smooth without constriction or unequal wall thickness. Practice making bends often, as this will improve timing control and therefore confidence.

Fig. 5.83.

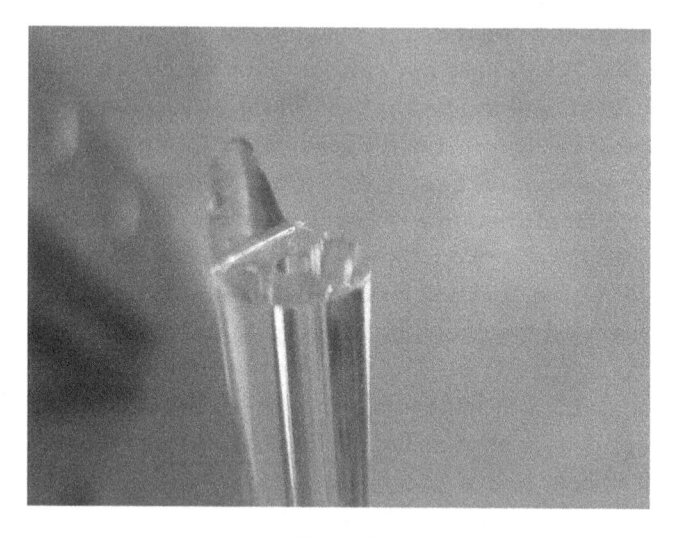

Fig. 5.84.

Select the capillary tubing and snap it in the normal way with the glass knife, the length will be a little longer than before at 450 mm. Great care must be taken when handling capillary tubing as quite often one may find a half moon lip of glass protruding from the cut surface (Fig. 5.84) using the back of the glass knife knock this off before proceeding to lightly flame polish

Fig. 5.85.

the end of the capillary tube, the small lip of glass is extremely sharp and may cause some sharp cuts to those who may not be paying attention. Mark the midpoint of the capillary tube, hold and rotate back and forth as previously practised. One will not have to "blow" the capillary tubing while making a bend which will make the exercise far easier.

Therefore, after making the first bend return to the flame which is now aimed to the right of the bend, swing the left hand in an arc in a similar manner as practised when making the second T piece (Fig. 5.85) with the right hand holding the glass on a central axis. The left edge of the flame should be positioned first 12 mm to the right of the first bend (Fig. 5.86), rotate back and forth until the glass softens, withdraw from the flame, hold the glass in the left hand horizontal but facing away (Fig. 5.87), lift the glass in the right hand to the horizontal position and in line with the first bend.

Allow to cool and inspect the work, there should be two in line 90° bends — the glass should still be able to be realigned slightly if necessary.

Return to the flame to the right of the second bend, the left edge of the flame which will again be 12 mm to the right of the second bend. The glass will now be offset but in line much like an engine crankshaft (Fig. 5.88) and in line with the previous bends as the right hand raises to create a third in line bend, allow to cool completely (Fig. 5.89).

Fig. 5.86.

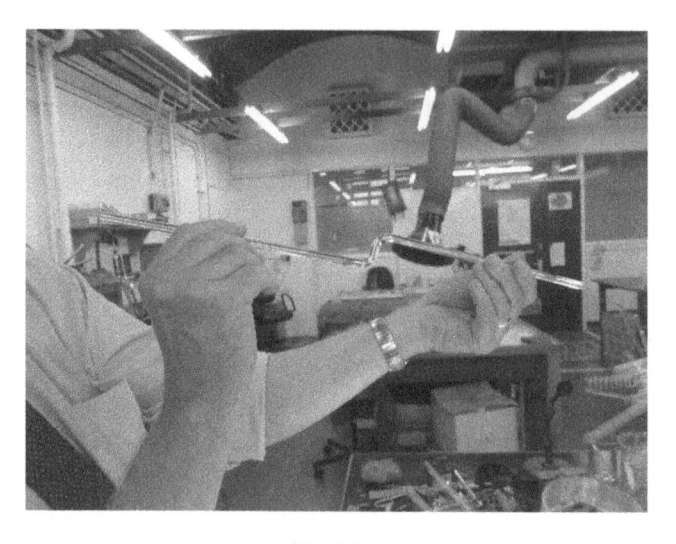

Fig. 5.87.

Once cool hold the bends in the left hand, once more play the flame to the right of the first bend, sweep the left hand in an arc while supporting on the central axis with the right hand until the glass can be felt to soften. The left hand holds the glass horizontally facing away but in line so that when the right hand lifts the glass up to the horizontal we have a fourth in line

Fig. 5.88.

Fig. 5.89.

bend — we have what is beginning to look like a set of stairs, return the glass to the flame to the right of the fourth bend when the glass has softened the left hand holds the glass vertically and the right hand raises the glass to the horizontal thus making yet another right angle bend.

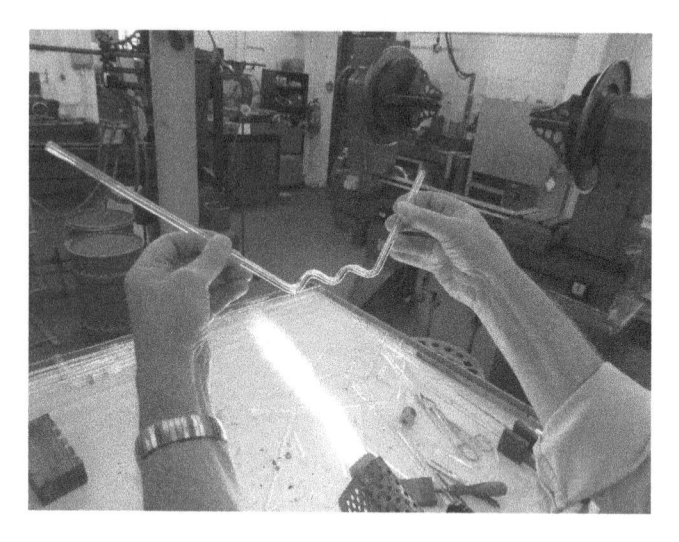

88 *Laboratory Scientific Glassblowing: A Practical Training Method*

Fig. 5.90.

As we progressively make more and more in line right angles remember that the left hand does all the rotating in whatever configuration and holds the glass steady while the right hand follows the rotation and lifts the glass to make the bend (Fig. 5.90).

Designate each hand to these specific tasks as confusion will reign once we pass the third bend, but so long as when the right hand is in a position to make the bend in line all will be well.

Let us now return to the tubing, prepare cut lengths of 8 mm ø tube 450 mm long and flame polish the ends. Seal one end with a cork, rubber cap or multipurpose tack. If one remembers the line: "cork to the ceiling, make the bend, and blow". Each time one makes a right angle bend while bearing in mind the configuration of the glass may be unusual, so long as the portion we are about to bend is held vertical, before making the sweeping bends towards one's self. "The cork to the ceiling, make the bend, and blow" will eliminate a great deal of mystery.

Practice making a series of "steps" using the various diameter tubing.

We have in a previous section made unequal joins with a domed or test tube end from 15 mm to 8 mm. Take one of the items while holding the 8 mm ø tube and gently warm the 15 mm ø body in a soft feather-like flame to avoid too sudden a thermal shock to the glass when the glass is warmed through, reduce the size of the gas, oxygen flame to a width of 5 mm ø.

Aim the tip of the flame midpoint along the 15 mm wide body when it can be seen to be melting withdraw from the flame and blow to produce a dome, aim the flame at the top of the dome, blow to produce a hole using the technique as in the T piece exercise, control the size of the hole to coincide with diameter of the "side arm" we are about to attach to be 8 mm ø 100 mm long (Fig. 5.91).

Heat both the edges of the hole and the tube until molten, withdraw from the flame and touch the tube against the hole to seal. Using either of the two "T" piece methods to "run in" the join. Once the "side arm" is joined increase the size of the flame to create a feather-like flame suitable to make the bend. The left side of the flame should be next to where the "side arm" is attached. Rotate back and forth in the flame. Withdraw from the flame, hold the body horizontally and with the 8 mm tube facing away from you, the right hand makes the bend, and then blow to produce a 90° bend coming straight off the body (Fig. 5.92).

Remove the cork, reduce the flame to give a sharp flame 5 mm wide, play the flame on the side arm level with the top of the 15 mm ø domed body. As the glass softens and melts pull the right hand away and burn through the spear point produced. Incline the glass as the flame is aimed at the sealed end — withdraw from the flame and blow to produce a small domed or test tube end. Readjust the flame to the feather type and warm the body and side

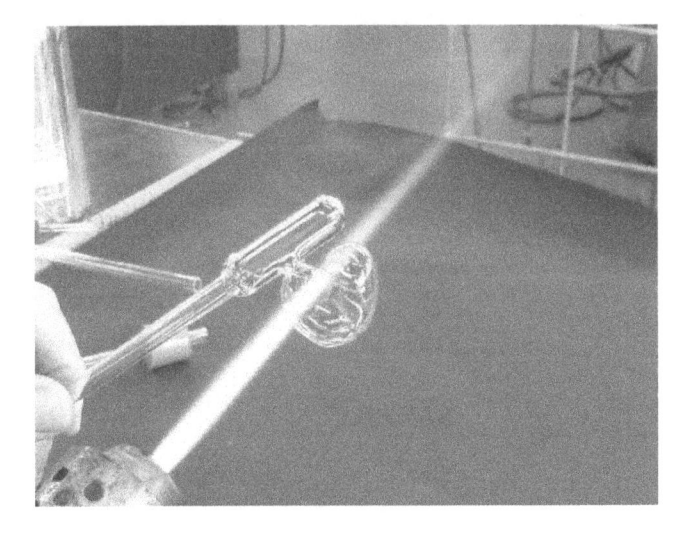

Fig. 5.91.

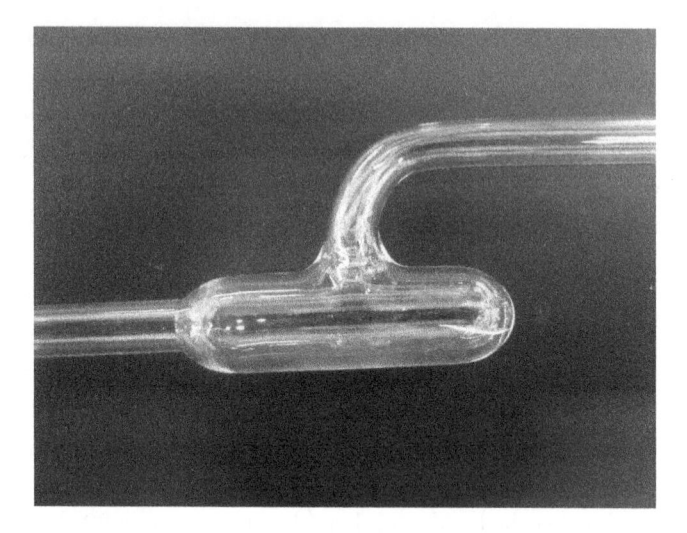

90 *Laboratory Scientific Glassblowing: A Practical Training Method*

Fig. 5.92.

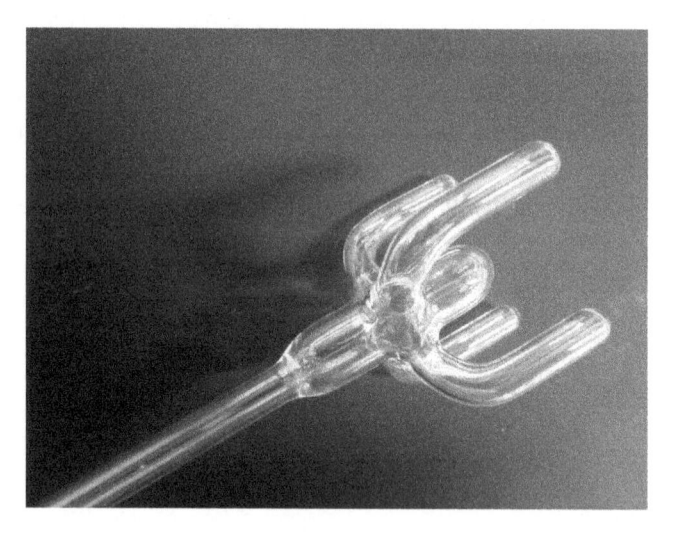

Fig. 5.93.

arm in the very top of the flame, which will dissipate any stain within the glass (flame annealing).

Repeat this exercise on the opposite side of the body and once more at right angles and finally opposite the third side arm to produce what looks like a cactus (Fig. 5.93).

5.8 U Bends

Select a length of 8 mm ø medium walled tubing and cut to 300 mm lengths and flame polish the ends.

Seal one end with either a cork or rubber cap and place in the left hand. "U" bends are best made while standing. The left hand will not be dominant. With both hands underneath the glass rotate the glass back and forth as practised when making the right angle bend (Fig. 5.94).

Use a wide feather-like flame, work at the top of the flame until one feels the glass soften. Withdraw the glass from the flame while still horizontal rotate for a further count of two. Stop rotating and start to make the bend by bringing both hands to the vertical (Fig. 5.95) in a graceful (Fig. 5.96), sweeping action until both tubes are parallel (Fig. 5.97). Hold this position, lean over towards the glass and blow down the open tube (Fig. 5.98), watch the tubing expand to its original diameter.

It is important that while blowing the tube is held vertically when the softened glass drops into the "U". Do not be tempted to hold the two parallel tubes horizontally as the "U" itself will sag out of line, which will then require unnecessary reworking.

Build confidence by practising with a variety of tubing diameters and wall thicknesses. Using an increased wall thickness leads to less likelihood of

Fig. 5.94.

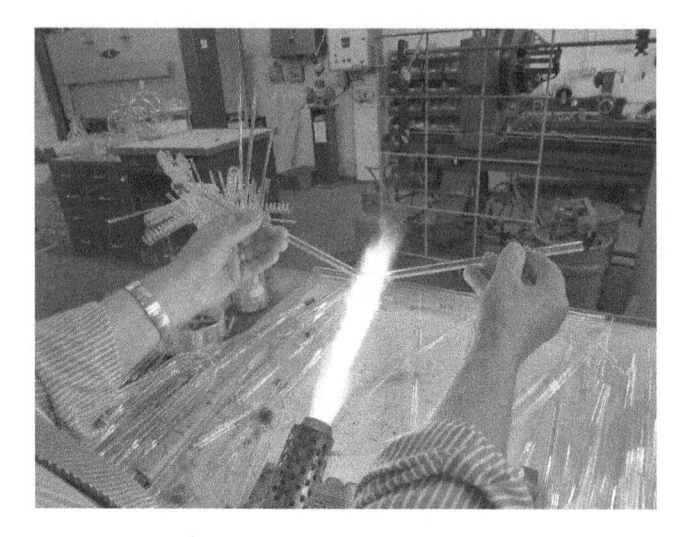

92 *Laboratory Scientific Glassblowing: A Practical Training Method*

Fig. 5.95.

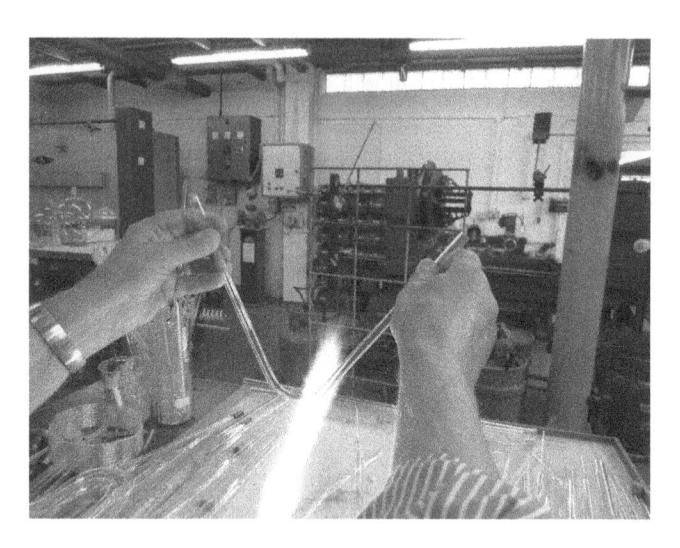

Fig. 5.96.

the bend collapsing, so much so that capillary tubing can be bent to shape without the need to blow. Therefore, practice with both capillary and rod will be a good exercise in making a nice graceful bend, where one can practise the timing of the sweeping hand motion and the need to hold still after the bend is made as there will be more residual heat within both rod and capillary and so a greater chance of the U bend sagging out of line.

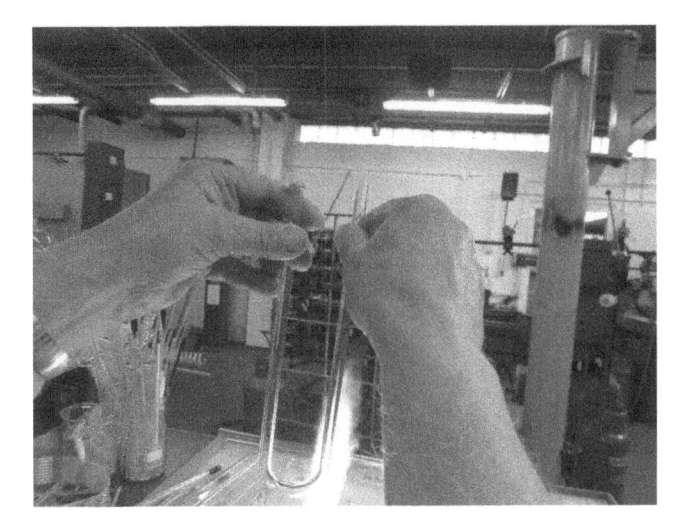

Fig. 5.97.

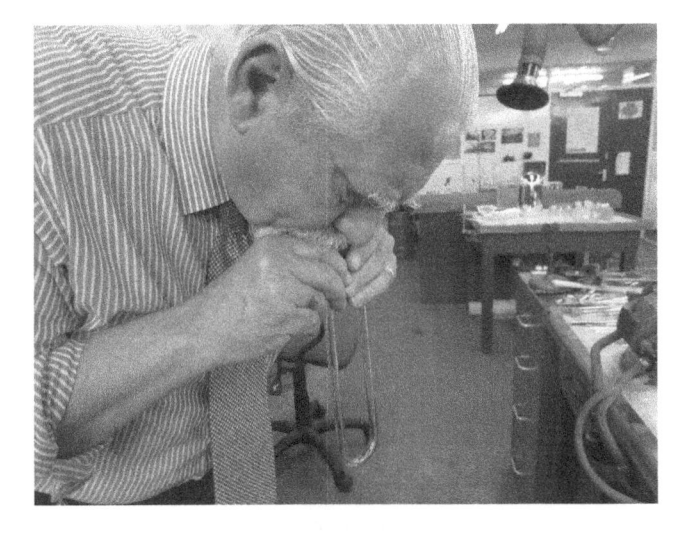

Fig. 5.98.

One may have access to a ribbon burner, if so this is an ideal burner for the making of U bends, especially the type that can be adjusted to give a variety of flame lengths (Fig. 5.99). The flame produced with this type of burner gives an even lateral flame, which in turn will melt the glass evenly and so should be reflected in one's ability to produce smooth even, wider bends. The technique is the same in that one will have to stand while both

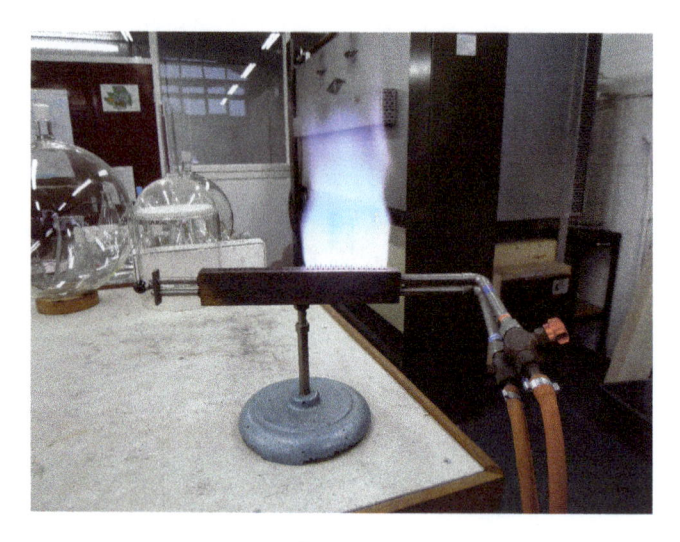

94 *Laboratory Scientific Glassblowing: A Practical Training Method*

Fig. 5.99.

Fig. 5.100.

hands hold the glass from underneath while rotating back and forth. The glass melts quite quickly so smooth rotation at the top of the flame is essential (Fig. 5.100).

One will feel the glass soften while both hands move together in a sweeping arc to hold the two side tubes vertically. Check the orientation both vertically and horizontally on either side of the bend as the glass solidifies. One can estimate the length of flame needed against a full scale drawing — measure the outer length of the bend and readjust after completing a number of trials and finally make a note of the length of flame to give a specific size of U bend for future use.

When a U bend has to be a specific width or length, the student may consider making the "U" from a manageable length of tube allowing it to cool then cut and flame polish the "U" 30 mm along the straight tubes either side of the "U" (Fig. 5.101).

The long pre-cut flame polished vertical tubes can be joined to the shortened "U" with flexible plastic tubing (Fig. 5.102). The manometer we have now produced would need mounting upon a backboard complete with a suitable scale (Fig. 5.103).

If the "U" requires reworking due to a constriction, misalignment if having been "overblown" which may well happen, a simple clamp or two will give support and rigidity as the bend is reworked (Fig. 5.104). This option may well be considered defeatist and the bend practised until mastered. Although on some occasions the complete bend is required to fit within certain dimensions, one can use a clamp as per diagram (Fig. 5.105) with holes to take the tubing set at the required distance.

When no gap between the verticals is required the U bend is made in the usual way, then each tube in turn is heated near the bend very

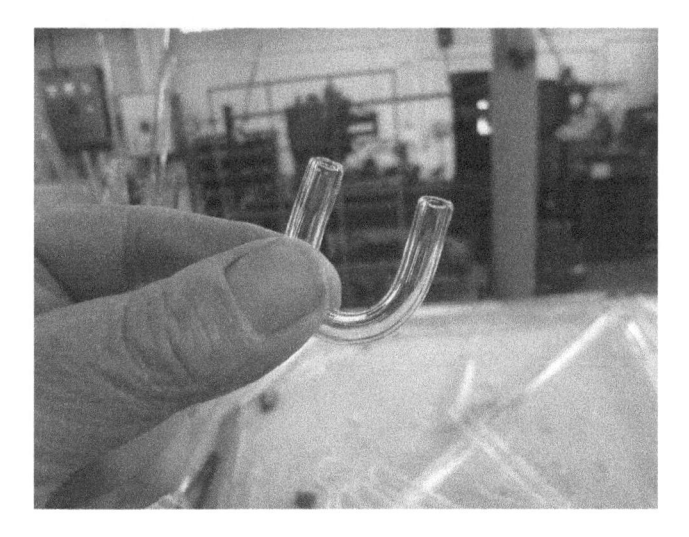

Fig. 5.101.

96 *Laboratory Scientific Glassblowing: A Practical Training Method*

Fig. 5.102.

Fig. 5.103.

Hand Techniques 97

Fig. 5.104.

Fig. 5.105.

gently and gradually drawn together (Fig. 5.106). This will require good directional flame control without overheating the glass which is only softened so that there is a noticeable resistance as the tubing is pulled in line (Fig. 5.107).

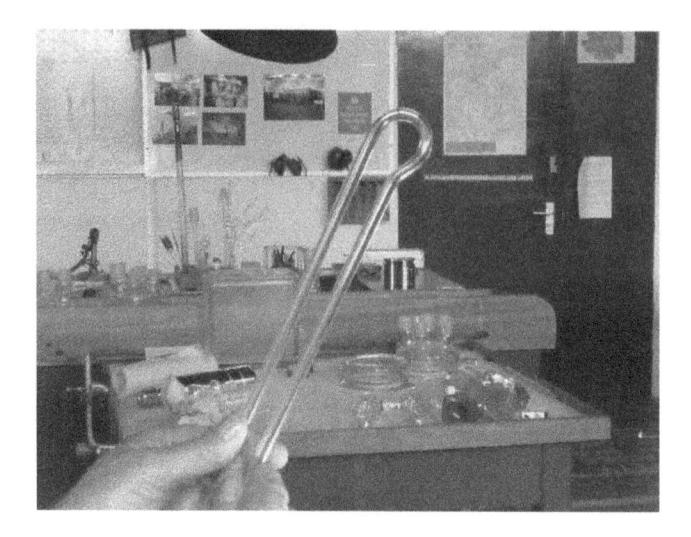

98 *Laboratory Scientific Glassblowing: A Practical Training Method*

Fig. 5.106.

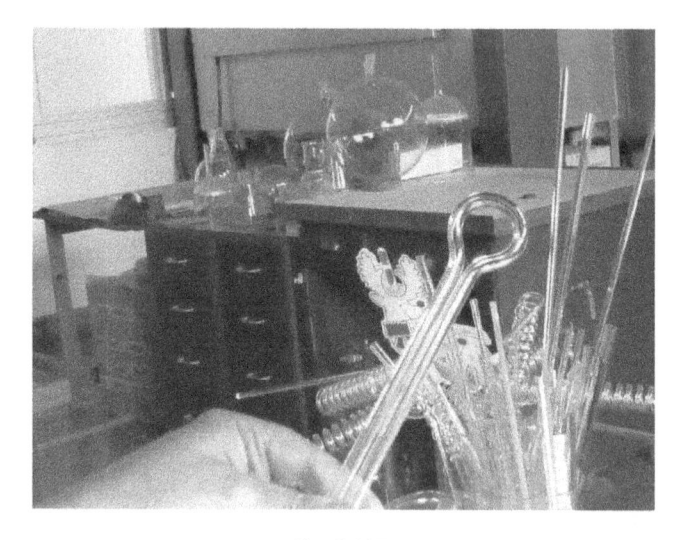

Fig. 5.107.

Where long lengths of tubing are used to make a tall U bend manometer an option is to make a 90° bend on each tube which is then cut to produce a short "stub" at right angle to the long tube, the edges of which are heated and sealed together two clamps are attached and tightened to give support as the bend is reworked to complete the "U" shape (Fig. 5.108).

Fig. 5.108.

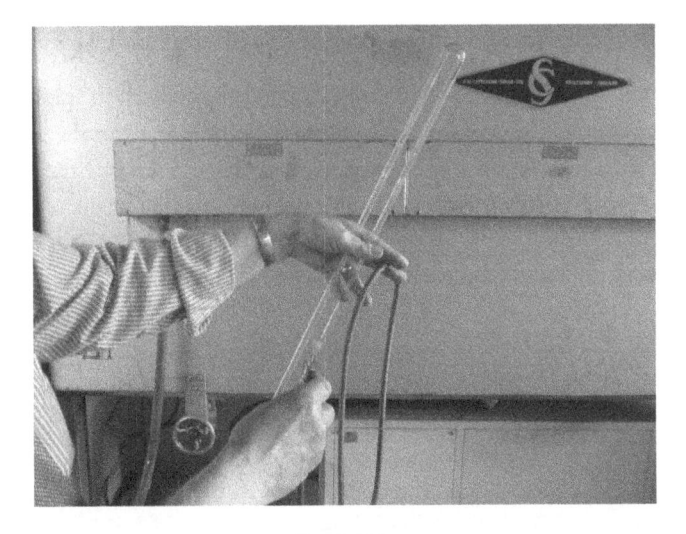

Fig. 5.109.

Should the verticals be very long, plus 1 m for example, a blowing tube attached to one of the verticals while the opposite tube is sealed off will enable the whole process to be much more manageable (Fig. 5.109).

We will return to the idea of supporting awkward seals when covering the use of the hand torch in Chapter 6.

5.9 Button Seal

Glass required   20 mm ø 400 mm long
7 mm ø 200 mm long

In Section 5.2, we have seen the technique of not only pulling a spear point but also how a domed end is made.

It would be of great advantage if the student practised pulling a number of spear points on the larger 20 mm ø tubing. The aim is to produce two strong straight points leaving 50 mm of straight tube between the shoulders. Every effort should be made to ensure that the points are on a central axis.

The method is similar to the previous section but the width of the flame will be 30 mm wide. In Section 5.4, we have previously pulled spear points from 15 mm ø tubing, the move up to 20 mm should not be too daunting.

Note: Try to keep the glass in the flame for a count of 15 or 20, withdraw from the flame, rotate for a further count of 10 before pulling the points.

Select the straighter of the two points, open the end with the glass knife and V block as previously described. This will now become the handle to be held in the left hand, aim a flame 12–15 mm wide at the shoulder of the right hand spear point while rotating the glass being held at an angle of 30° (Fig. 5.110).

Fig. 5.110.

Fig. 5.111.

Once the point is removed continue to heat the shoulder, withdraw from the flame and blow to produce a domed end, readjust the burner to give a fierce small 5 mm wide flame. Heat the very centre of the dome (whilst still rotating), once hot withdraw and blow to make a thin bubble which is removed (Fig. 5.111).

The hole required will be 10 mm diameter. If the hole is too small we can enlarge it by "reaming" the hole with either a sharpened/faceted piece of graphite rod or a brass reamer lubricated with graphite/beeswax. Using a 10 mm wide flame, rotate the edge of the hole in the edge of the flame until starting to soften, insert the carbon rod or reamer and hold rigidly in the right hand as the left hand rotates as the softened hold is rested against the rod or reamer the hole should open gradually (Fig. 5.112).

Repeat the process until the desired diameter is reached. We will see that the "button" we are about to produce is 10 mm diameter; the hole ideally should be 10 mm diameter so that the top part of the button rests on the ridge of the hole. Warm the end of the dome in a feather-like flame to distribute the heat and put to one side.

We will now prepare a button on the 7 mm ø tube, take one of the pre-cut and glazed pieces and mark the centre point, using the top of a very small 5 mm ø flame. Aim at the mark, rotate holding the glass — left hand "over" right hand underneath. The left hand is dominant concentrating on rotation

102 *Laboratory Scientific Glassblowing: A Practical Training Method*

Fig. 5.112.

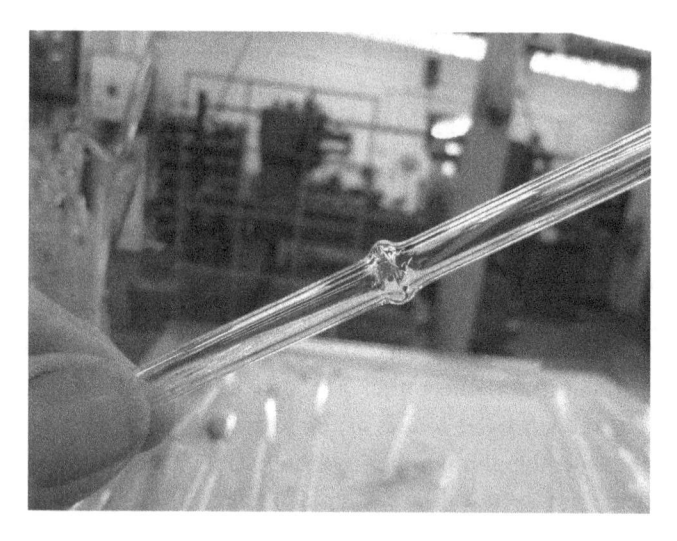

Fig. 5.113.

the right hand following and keeping the glass central, as the glass softens gently push and turn with the right hand to form a noticeable ridge or button (Fig. 5.113), on the 7 mm ø tube the button will be approximately 10 mm ø — withdraw from the flame keep rotating and allow to cool. The button should not be pushed together until it becomes flattened against itself, but be a hollow ridge.

Fig. 5.114.

Once cool mark 40 mm to the right of the button — cut and flame polish.

Adjust the burner to give a flame 5 mm wide, seal off the end of the spear point. Hold the spear point in the left hand and slide the 7 mm tube into the prepared hole and rest the button on the ridge of the hold (Fig. 5.114). Rotate and concentrate the flame on the shoulder and button as the glass softens, push together slightly. At this point withdraw from the flame and blow to see that the seal is free from leaks, return to the flame, watch the seal between the button and the shoulder as the glass is heated once more, withdraw once more and blow. One should see that a dark ring has formed at the seal as the inner tube has been "run in" allowing one to see end-on into the inner tube (Fig. 5.115). At this point we must concentrate upon the shoulder — only the right-hand side of the flame plays upon the seal as the majority of the flame plays on the shoulder as the glass is rotated and held at a 45° angle. The molten glass will drop into the shoulder as we found when making an unequal join (Fig. 5.116).

As we withdraw from the flame drop the left hand down to the vertical there will be sufficient glass to form a strong, robust hemispherical shape as we blow (Fig. 5.117). Watch as the shoulder is formed, once satisfied bring the glass to the horizontal while still rotating. The unsupported inner tube

Fig. 5.115.

Fig. 5.116.

will still be "moving", balance the rotation and stop if necessary to allow the inner tube to centralise. If unsuccessful first time the seal/shoulder area may be warmed and as the inner tube is seen to move, remove the glass from the flame and try once more to get it to sag into the central axis. Allow to cool and anneal the piece. This completes the first stage of the exercise.

Fig. 5.117.

Fig. 5.118.

Repeat the first exercise. After first preparing a second piece of 7 mm ø tube 150 mm long we will prepare one end of this tube by heating one end in a 15 mm wide flame. Once softened insert the carbon reamer and hold rigidly in the right hand, rotate and incline the left hand to form a slight "flare" to the end of the tube (Fig. 5.118). This will become the "side arm". Allow to cool and place to one side.

Continue to hold the spear point in the left hand having already made the seal, formed the shoulder and centralised the inner tube. Introduce air into the flame to produce a feather edged flame in preparation to make a bend. Heat the 7 mm tube to the right of the seal, the left edge of the flame being 5 mm to the right of the seal, heat the tubing until it can be felt to start softening — withdraw from the flame, raise the left hand, hold the glass vertically and complete the bend by sweeping the right hand towards you in a graceful arc and blow the bend while holding the position (Fig. 5.119).

Reduce the size of the flame to 5 mm wide, using the tip of which aim at the 20 mm ø tube just below the shoulder opposite the newly made bend (Fig. 5.120) — withdraw, blow to form a bulge; return to the flame and aim once more at the top of the bulge, withdraw and blow a small thin bubble (Fig. 5.121) which when removed will reveal a hole. Seal off the 7 mm tubing with a rubber cap, bung or tack. Increase the flame to 10 mm wide and whilst the glass is still hot aim the flame across the face of the hole, at the same time heat the edge of the flared side arm, once both are hot withdraw from the flame and bring the two edges together (Fig. 5.122).

"Touch–Pull–Blow"

Run the join in using a combination of both methods used in Section 5.5 — once complete withdraw from the flame and adjust and soften the flame in preparation for a second bend.

Fig. 5.119.

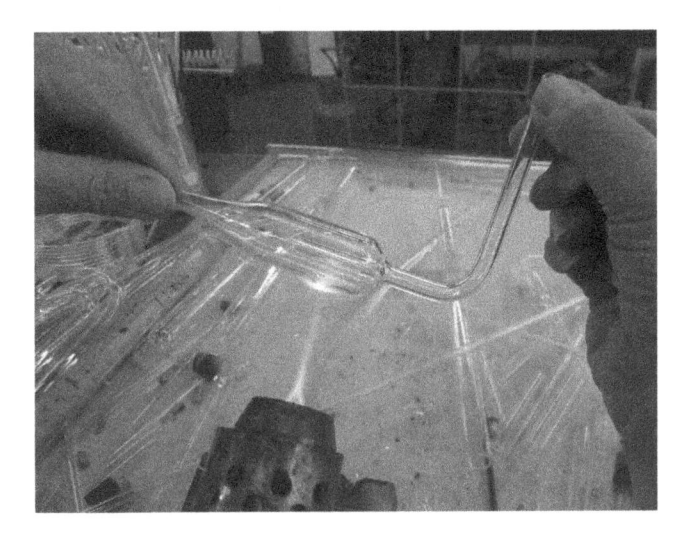

Fig. 5.120.

Fig. 5.121.

Heat the side arm — rotate the left hand back and forth in an arc while the right hand supports the side arm on the central axis. Once the side arm starts to soften — withdraw, hold the body horizontally with the seal facing and the first bend facing vertically, sweep the right hand towards you, make the bend and blow (Fig. 5.123). The first and second bends should be in line,

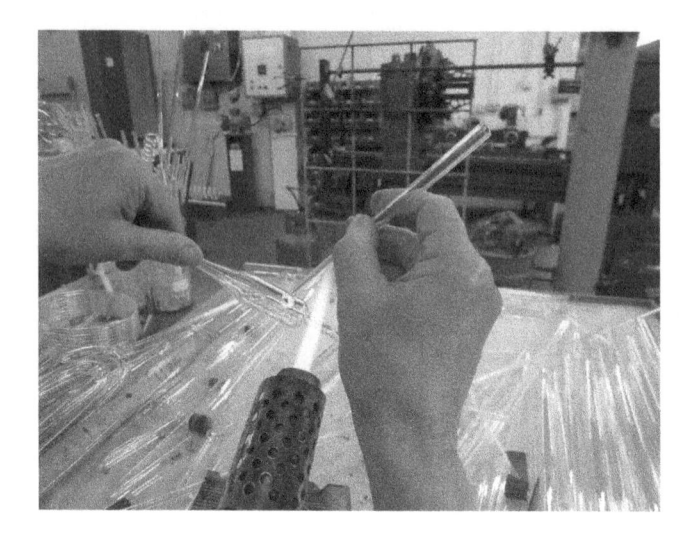

108   *Laboratory Scientific Glassblowing: A Practical Training Method*

Fig. 5.122.

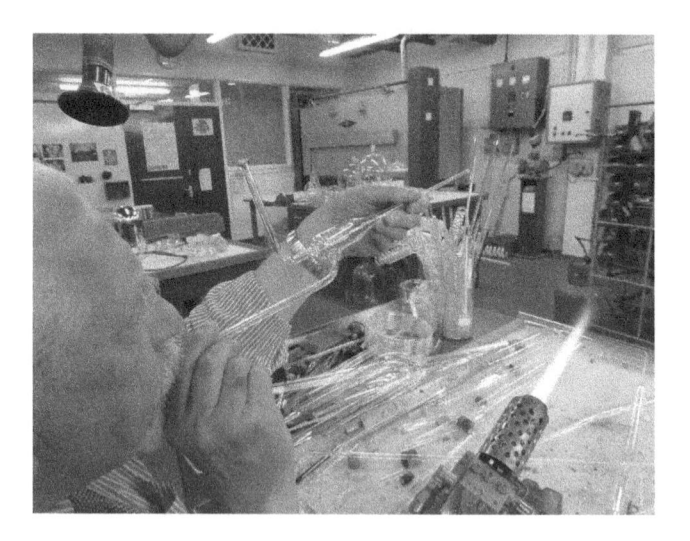

Fig. 5.123.

allow to cool slightly before aiming the flame to the right of this second bend, heat evenly and when softened withdraw the glass from the flame — hold the glass vertically. The second bend facing, sweep the right hand upwards and blow to produce a third right angle bend which should be in line with and at the same height as the first bend (Fig. 5.124).

Fig. 5.124.

Fig. 5.125.

Allow to cool, once handleable scribe a mark on both of the arms 40 mm from the bends, rest the glass on a V block inline and opposite the mark, moisten the mark and tap the excess with the back of the glass knife. The glass will break at this point — flame polish the ends.

This exercise is based upon a Dreschel Head (Fig. 5.125).

Fig. 5.126.

5.10 Supported Internal Seal

Unsupported seals are adequate for light smaller internals. Longer heavier internal tubular multi-walled glassware support for the internal tubing is required. There are a number of ways of supporting the internal tube. The simplest of which is corrugated card, simply because air will be able to pass along its length due to the linear corrugations, this is an important consideration for whatever support material we may use.

Corrugated card is relatively cheap, easily cut to shape with scissors and does not expand unduly (Fig. 5.126). The only drawback is that it tends to burn if too close to the working area, when alternative support material will have to be sought.

When an annular space is required sufficient corrugated card can be wrapped around the internal tube to make up the gap. A few points to note are: The corrugated card must first be flattened a little by rubbing across the ridges with the back of a glass knife (Fig. 5.127), this will soften the ridges and enable the card to more easily take on the curve of the inner tube. The start and finish of the wrapping should be at the same position so as to avoid any offset or a misaligned inner (Fig. 5.128).

The card should be at least 60 mm from the working area, yet overhang the opposite end of the internal tube so as to enable the card to be gripped on completion of the seal for its removal (Fig. 5.129). It follows that the

Fig. 5.127.

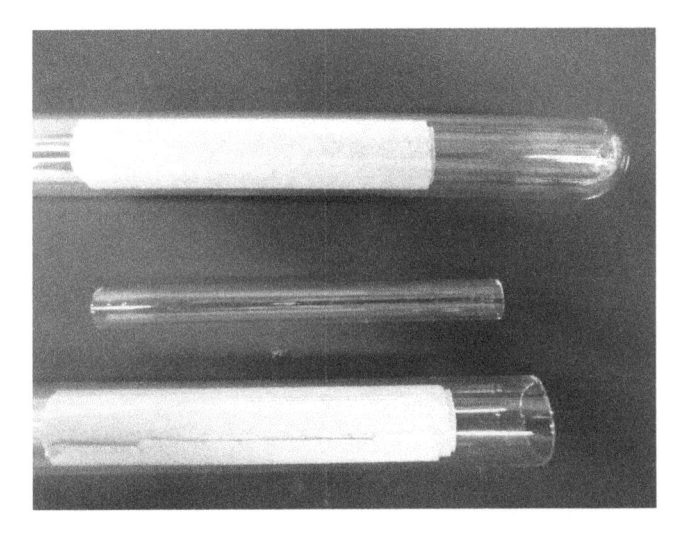

Fig. 5.128.

external tube will also have to extend beyond the support packing for it to be sealed with a cork or bung later in the process (Fig. 5.130).

The exercise we are about to commence is based upon a trap used in conjunction with vacuum systems therefore the seal must be free from leaks. This type of trap is made in a variety of lengths and diameters.

Fig. 5.129.

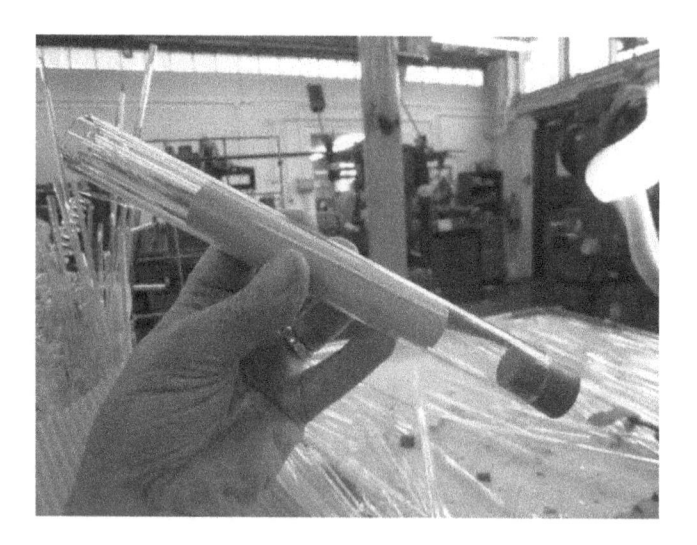

Fig. 5.130.

Prepare the tubing as follows:

Inner tube 14 mm ø 130 mm long
Side arm 14 mm ø 170 mm long
Inline tube 14 mm ø 170 mm long

Fig. 5.131.

Large outer tube 26 mm ø
Method one: domed end 26 mm ø 270 mm long
Method two: 26 mm ø 230 mm long
The unit will use medium-walled tubing throughout
Corrugated card after flattening
110 mm × 230 mm
The 14 mm ø tube can be snap cut as described in Section 5.1
The 26 mm ø tube is best cut by "hot point" as described in Section 5.1

To form the domed end we must first "draw down" the 26 mm tube. Adjust the flame to give an oxygen enriched flame 20 mm wide. Place the tube in the left hand while holding a piece of approximately 7 mm ø rod in the right, heat the very end of the tubing until it starts to collapse while at the same time heat the end of the rod to redness — withdraw from the flame and touch together and centralise both tube and rod — move the flame to the left of this initial join by 20–30 mm. Heat the tubing until soft and pull a spear point, burn off the rod and excess spear leaving 150 mm to hold once cool. Incline the glass and remove the spear at the shoulder to start making the domed end. On larger diameters three or four indentations in the end of the softened tube with a carbon rod will make it easier to attach the glass rod (Fig. 5.131).

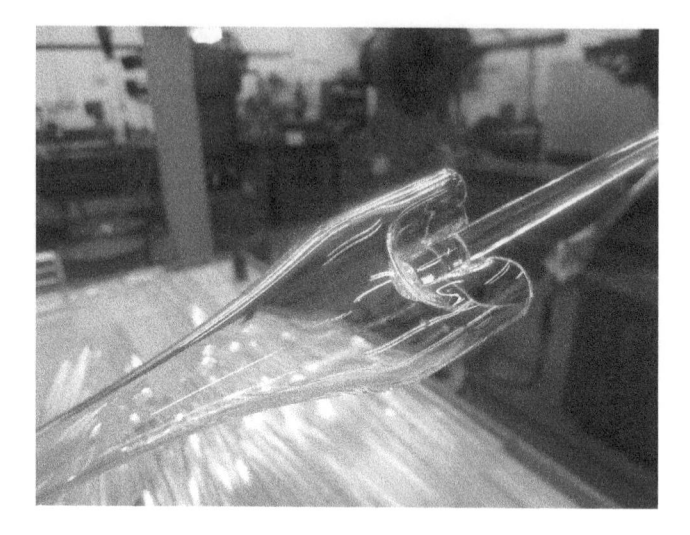

114 *Laboratory Scientific Glassblowing: A Practical Training Method*

Fig. 5.132.

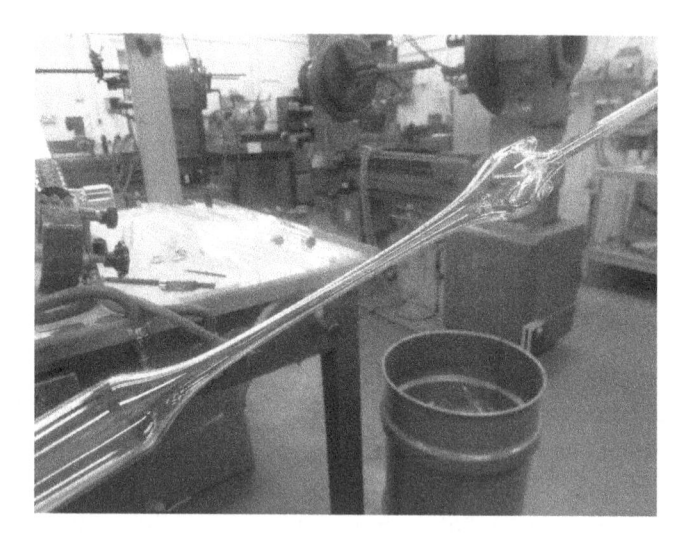

Fig. 5.133.

A third method to try is to heat one side only of the 26 mm ø tube close to the opening, once molten lay the rod which is not softened but still quite rigid onto the molten glass but parallel to the tube (Fig. 5.132), push the rod in so that it lies centrally and on axis with the larger tube as the glass starts to solidify, move the flame to the left and pull a spear (Fig. 5.133) then proceed to the domed end (Fig. 5.134).

Fig. 5.134.

Fig. 5.135.

The Exercise

The inner tube is supported and lies centrally within the outer. The inner tube is positioned just touching the inside of the domed end (Fig 5.135).

Warm the dome in a feather-like flame and once warm reduce the size of the flame but increase the intensity, aim the flame across the dome, the end of which should start to flatten and make contact with the inner tube, incline

Fig. 5.136.

the glass in the flame and watch the seal forming as a contact line starts to show, withdraw from the flame and blow, inspect the end of the dome and the seal which should be an even thickness and colour (Fig. 5.136).

Return the glass to the flame and concentrate the flame on the end of the inner tube, once molten withdraw and pause while still rotating, then blow to produce a thin bubble which is then removed by slicing off with the back of the glass knife. Place a cork or bung in the open end of the larger tube to seal. Take a prepared piece of tubing the same diameter as the inner and heat both the open seal end and the yet to be joined tube in the edge of the flame, once hot withdraw, touch together, slight pull and blow (Fig. 5.137). Once satisfied that the seal is made return to the flame which should be aimed at the shoulder, only the right hand edge of the flame should be on the seal, imagine that one is making an unequal join and are about to make a rounded shoulder — withdraw from the flame, drop the left hand, lean over the glass and blow the glass to size. Watch as the shoulder is formed and the seal is confirmed. The shoulder/seal may require some more work, try to keep this to a minimum. Proceed to blow a hole for the sidearm just below the shoulder, once the hole has been made, treat the next action as if it were a "T" piece.

Take both the body and side tube and position their edges on either side of the flame (which should be the same width as the hole and side arm).

Fig. 5.137.

Fig. 5.138.

1: Edges hot — 2: Withdraw from the flame — 3: Touch edges together — 4: Slight pull — 5: Blow (Fig. 5.138).

Reduce the width of the flame to 5 mm wide and proceed to "run in" the side join section by section. When one is fully confident a 360° back and

Fig. 5.139.

forth movement will complement the sectional method. Make sure that the side arm is at 90° right angle and in line with the body when viewed end on. Warm the whole seal and side arm areas with a feather-like flame until small flecks of orange flame can be seen, which will indicate that the temperature is balanced. Once cool remove the packing. Seal the side tube with a cork while holding the inline tube in the left hand proceed to "draw down" the end of the larger body, seal off and produce a domed end 150 mm from the seal without it touching the inner tube. To complete the exercise cut the side arms to length and glaze (Fig. 5.139).

A point to note: The right hand position, we have until now used the V formed by the first and second fingers in combination with the thumb to rotate the glass. With larger/heavier diameter tube being held and supported in the right, the V should be formed between the second and third fingers in combination with the thumb (Fig. 5.140). The third and fourth fingers do not move while the first and second fingers "click" the tubing against the thumb to rotate the glass.

An alternative method from the domed end start of the supported seal is to prepare the internal tube centrally leaving 60 mm to the right of the inner tube, heat the edge of the larger, make three or four indentations if felt

Fig. 5.140.

Fig. 5.141.

necessary and attach (tack) rod to form the handle, move the flame to the left in line with the inner tube — heat, rotate and allow the glass to constrict until it touches the inner, move the right hand away as if making a spear point and burn off at the shoulder (Fig. 5.141), "blow" to complete the seal. The rod

120 *Laboratory Scientific Glassblowing: A Practical Training Method*

Fig. 5.142.

may be attached centrally after the glass has been reheated and drawn away to remove any excess thickness of glass that may have gathered at the end. Reheat the end of the inner tube and blow out, then proceed once more to join a central tube and side arm.

We will now attempt yet another method of making the "seal" from the point at which we have "blown out" the centre of the inner tube, having first prepared the tube we are about to attach by "flaring" it slightly with a carbon reamer (Fig. 5.142). Rotate the very edge of the opened seal in the edge of a hot fierce flame until quite molten. The glass will form a rounded edge due to surface tension, likewise the flared edge of the smaller tube should be about to collapse! Bring the two edges together and blow (Fig. 5.143). This is very much dependent on timing and the use of a hot flame, it is a technique that can be combined with the two previous methods in future work.

Practice the exercises in progressively larger diameters:

Suggested sizes: D1 26 mm   D2 14 mm   L1 150 mm   L2 130 mm
                  D1 32 mm   D2 15 mm   L1 200 mm   L2 150 mm
                  D1 38 mm   D2 16 mm   L1 220 mm   L2 180 mm

Fig. 5.143.

Bear in mind that the above dimensions are a final size — make allowance for the D1 and L1 when making the unit from a domed end start or open end start.

The length and width of the corrugated card will alter to accommodate the changes in annular space.

5.11 Items 1–9 Repeated in Progressive Larger Diameters

Spear Points

There are no alternatives to practising and making good, strong axially straight spear points! Some may consider that this attitude is very much "old school". Their importance cannot be overemphasised in understanding and feeling the effect of gravity on molten glass.

At first the student will have a natural desire to progress to apparently more spectacular and interesting items of glassware.

Whenever I pull a spear point I remember the 19¢ microbiologist Louis Pasteur, one of his many observations made him realise that at the biological level cross-contamination was occurring in part by the syringes of the day in that the seal between the plunger and the body was made by a leather washer which was difficult if not impossible to sterilise. His nephew Andrien Loir pulled tapers on glass tubes to a sharp point, in effect he was pulling spear points. The points were used to take samples as well as to inoculate! Used only once these Pasteur pipettes avoided the possibility of cross-contamination.

Should I ever tire of pulling spear points my motivation is to remember Pasteur. The pipette that bears his name and the millions of lives this simple drawn tube has saved.

The greater the diameter of tubing from which a spear point is to be made will affect both the right hand hold and the starting point.

The "hold" for the right hand on smaller diameters is from underneath with the glass resting between the first and second fingers while the thumb rotates the glass (Fig. 5.144) as the diameter increases the glass will have to be held between the second and third fingers with the first and fourth fingers in support. The thumb position is still in line with the index finger as it continues to give rotation to the glass (Fig. 5.145). The left hand remains the same with an overhand grip, the second, third and fourth fingers giving support and the first and thumb rotating.

If we take a piece of 30 mm ø tube of some 40 cm in length and mark the centre we will have to use a flame of similar diameter. Working in the top of a gas, air, oxygen flame one must expect that the time taken to melt the glass will be longer and so the amount of concentration required to control the molten glass will have to increase accordingly.

Fig. 5.144.

Fig. 5.145.

Fig. 5.146.

Overcome the urge to withdraw from the flame as soon as the glass starts to melt by rotating evenly and by keeping the hands together and not allowing them to move apart. Once satisfied the glass has sufficient heat withdraw from the flame, rotate evenly as you feel the glass starts to stiffen at this point, pull apart whilst continuing to rotate the glass (Fig. 5.146).

Fig. 5.147.

If starting a spear point from an open end, heat the end until just starting to soften, then make a series of indentations with a carbon rod to reduce the diameter sufficiently so that a piece of rod or tube may be "tacked" on the end to form a handle (Fig. 5.147). Move the flame past the indentations and proceed to pull a spear point from the main body of the tube. An alternative start from an open end is to heat one side only, then place the end of a piece of rod parallel to the tube and push it into the molten glass until it reaches the centre. After allowing it to cool slightly, proceed to pull the spear point from the main body (Fig. 5.148).

Straight Join

Using the methods previously described straight joins can be made on the bench by hand up to 30 or 40 mm ø with practice. From 40 mm ø the joins are best made after first pulling spear points as the grip of the right hand will be difficult on 30 mm ø plus diameters. Working with larger diameters require larger flames and as a result an increase in radiated heat, so once more spear points will enable the hands to be a little farther away from the flame (Fig. 5.149). Access to a glassblowing lathe will increase the scope for the trainee to handle larger diameter glass tubing and rod.

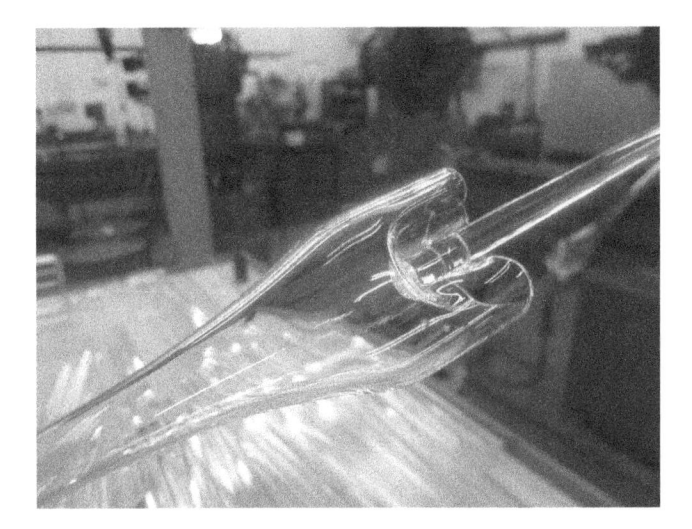

Fig. 5.148.

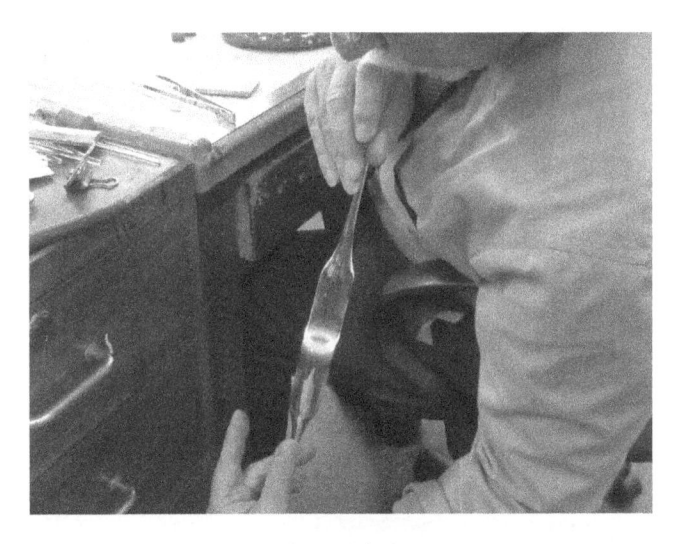

Fig. 5.149.

There is the worry that the trainee will become more dependent upon the lathe rather than attempting to manipulate the glass by hand.

The facilities available and the working environment can be deciding factors as to when the trainee moves onto using mechanical aids to complete a glassblowing task.

I have emphasised throughout the earlier sections of this manual the importance of understanding the effect that gravity has upon molten glass and the feel and control needed to make the glass conform to size, shape and wall thickness.

Scientific glassblowing has to conform to various specifications relating to design as well as any relevant safety considerations. The use of a lathe can and will make reproducibility easier to achieve by the very fact that the "inline" rotational movement of two pieces of glass that are to be joined are not solely dependent upon the individual's hand skills. There has to be a balance between both bench and lathe work so that the full potential of both the individual and material can flourish. If a glassblowing lathe is available it will be obvious that there are differences from a metal or woodworking lathe in that there are two "chucks" or "heads" which are synchronised to start and increase in speed, then slow down and stop in unison.

The "chucks" can range from engineering (Fig. 5.150), sun and planet (Fig. 5.151) or scroll (Fig. 5.152) which are opened by an external "scroll" ring and a key for the engineering and sun and planet or even automatic self-centralising spring loaded chucks (Fig. 5.153). The larger heavier lathes tend to rotate at lower speeds for use with larger diameter tubing whereas some smaller lathes can be termed "centrifuge" lathes where the molten glass not only overcomes gravitational forces but is spun outwards by centrifugal force often against carbon formers (Fig. 5.154).

Fig. 5.150.

Fig. 5.151.

Fig. 5.152.

The Exercise

Ensure that the lathe is switched off and that the speed control is set to zero. Take two pieces of 40 mm ø tubing 40 cm long. Take a piece and seal one end with a cork, this will be placed in the right hand chuck. Grip the glass

128 *Laboratory Scientific Glassblowing: A Practical Training Method*

Fig. 5.153.

Fig. 5.154.

lightly in the jaws of the chuck. Attach a length of rubber tubing via a rubber bung to the second 40 ø tube and place in the left hand chuck with the rubber tubing through the lathe head stock and secure it to a swivel and thence to a blowing tube (Fig. 5.155).

Fig. 5.155.

Once more ensure that the speed control is set at zero, switch the lathe on, gradually increase the speed of the lathe to just a little faster than one would be able to rotate by hand. At this point we have the option of using either a hand torch or burner attached to the lathe carriage. Ignite either burner and move the right hand "chuck" so that the edges of the tubing are close together, position the flame so that both edges of the glass are in contact with the flame, increase the flame temperature and watch as the edges melt. If using a hand torch raise it away from the molten edges and move the right hand chuck so that the edges touch together, then pull apart slightly and blow, return the flame to the join and reheat until molten once more, raise the flame away and using a controlled blow regain the original diameter.

Repeat as necessary to produce a neat "run in" join without ripples and loss of wall thickness. If using the lathe burner do not remove the flame but bring the two edges together while still in contact with the flame Touch–Pull–Blow. If necessary use a flat carbon paddle to blow the glass against. The carbon rests against the glass using the solid area either side of the molten join to give the correct diameter (Fig. 5.156). Switch off the flame and blow against the carbon as the glass cools down. We will expand on the use of lathes in Section 5.18. The comments are also relevant to unequal joins (see Sections 5.17 and 5.18).

Fig. 5.156.

Bends, 90° and U Larger Diameters

We have seen that to make a 90° bend we first must heat sufficient length of tubing to reach around the outer circumference of the bend using a wide feather-like flame and when we "feel" the glass softening the left hand holds the glass vertically with the cork directed at the ceiling as we make the bend by sweeping the glass towards you with the right hand. A blow at this stage is given while keeping the glass held at 90° as the glass increases in size to regain the original diameter.

Similarly with the "U" bend both hands working in unison as they sweep vertically in a graceful arc. It will soon be obvious that not all bends are either 90° right angles or U-shaped. They quite often have a variety of angles and diameters. Larger diameter bends require a combination of both methods when heating the tubing either along the length of a bench burner or a ribbon burner flame. As we feel the glass softening as well as watching an even sodium flame radiating from the glass surface (Fig. 5.157), withdraw from the flame and "drop" the hot glass into the bend by holding the sealed end of the tube in the left hand away from the body as the right hand holds the open end closer to the body, the softened glass sags under gravity into the bend as each hand holds the glass at 45° to the vertical. A constant pressurising blow will recover the tubing diameter (Fig. 5.158).

Fig. 5.157.

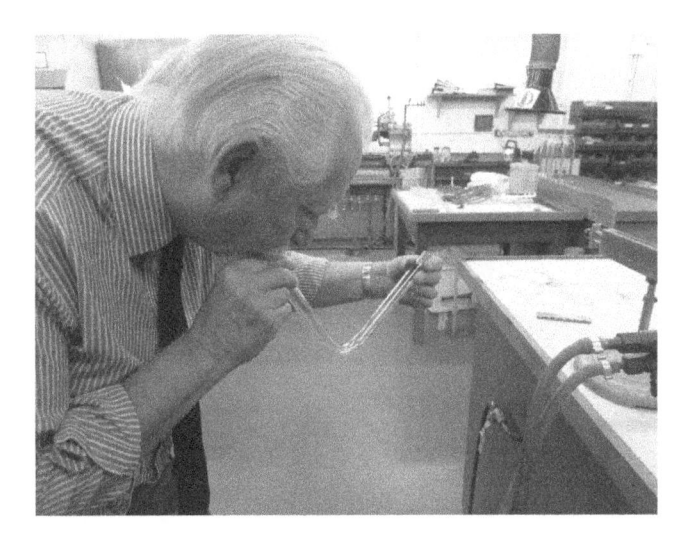

Fig. 5.158.

One may sense that both hands "flick" the hot glass into the bend as both hands move upwards to hold the 90° right angle. Very large diameter bends require specialised burners to ensure that the glass is heated evenly throughout the wall and the bend made gradually while the tubing diameter is maintained by constant internal pressure.

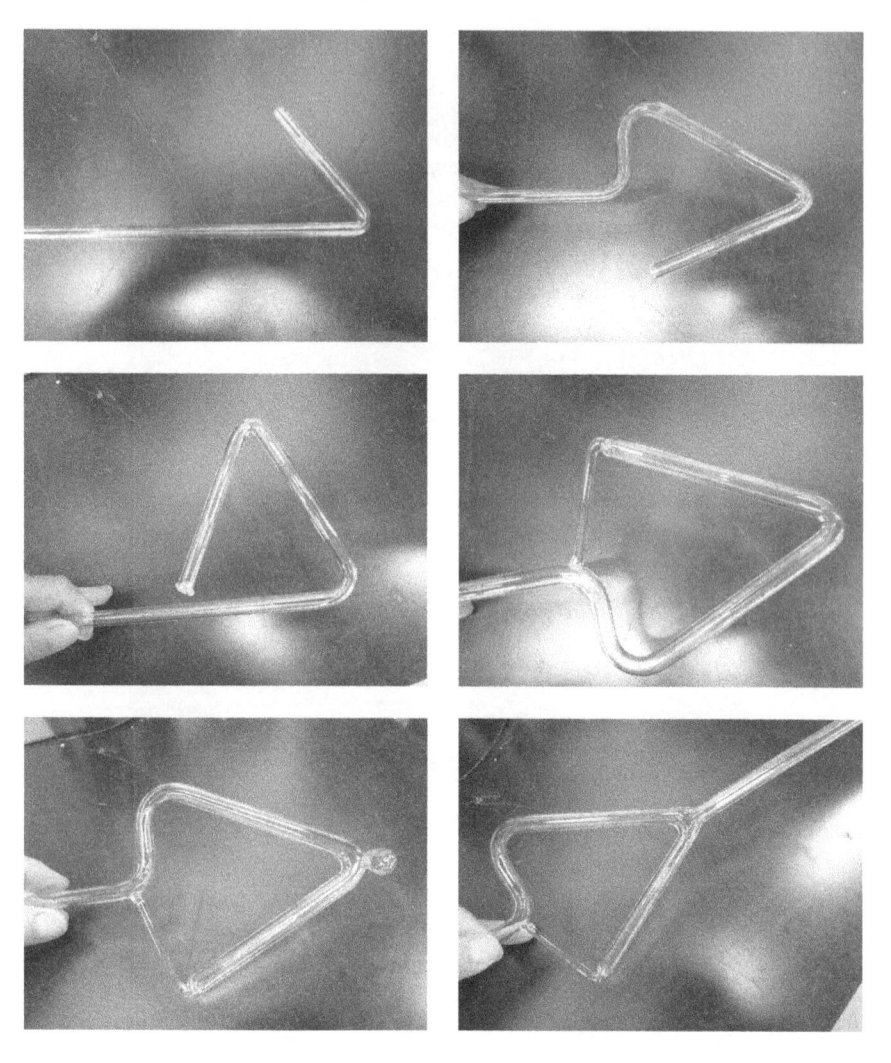

Fig. 5.159.

Y Pieces

Depending upon the diameter and length of a Y piece a problem arises occasionally that the whole of the seal area overheats and becomes misaligned, this can be caused by poor flame control or lack of wall thickness at the seal. To overcome the misalignment one can "brace" the seal by bringing one arm in line with the centre of the bend, the remaining arm is then sealed off and

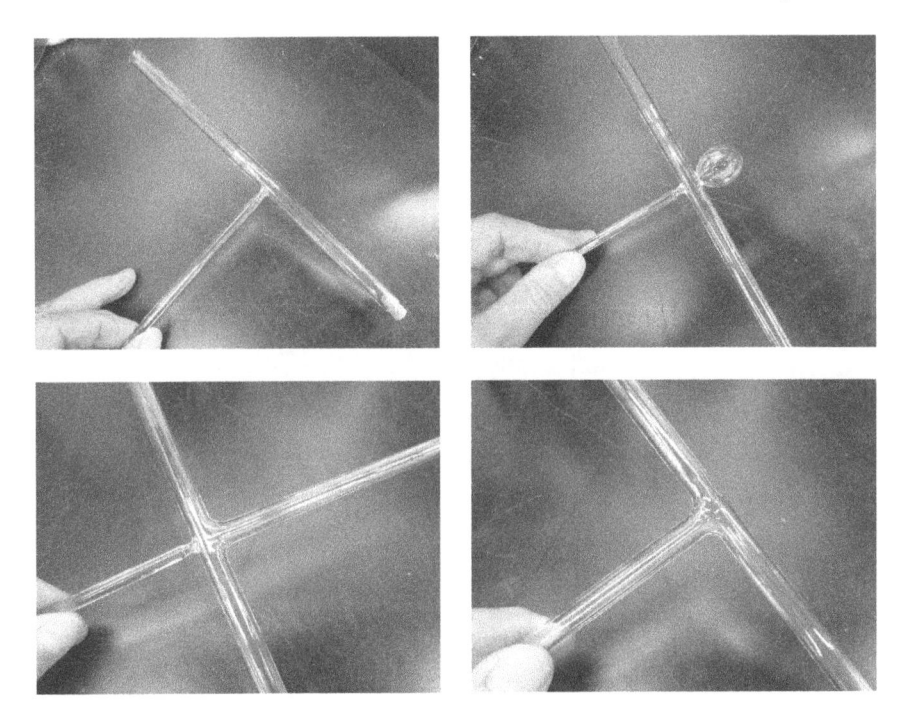

Fig. 5.160.

the excess glass used to "brace" or attach itself to the first side arm which will allow the join to be completed as if it were a simple straight join.

See photo sequence (Fig. 5.159).

T Pieces

Attaching a rod at 90° to a tube will simplify the making of a T piece by enabling the side arm to be attached as if a straight join. The rod is removed on completion of the join (Fig. 5.160). If time allows jigs can be fabricated to aid reproducibility. Bearing in mind that one should not diminish the skill needed in flame control by lack of practice of the basic method but use alternative methods to expand one's skills and use them when required and appropriate. Supported internal seals will be covered in Sections 5.17 and 5.18.

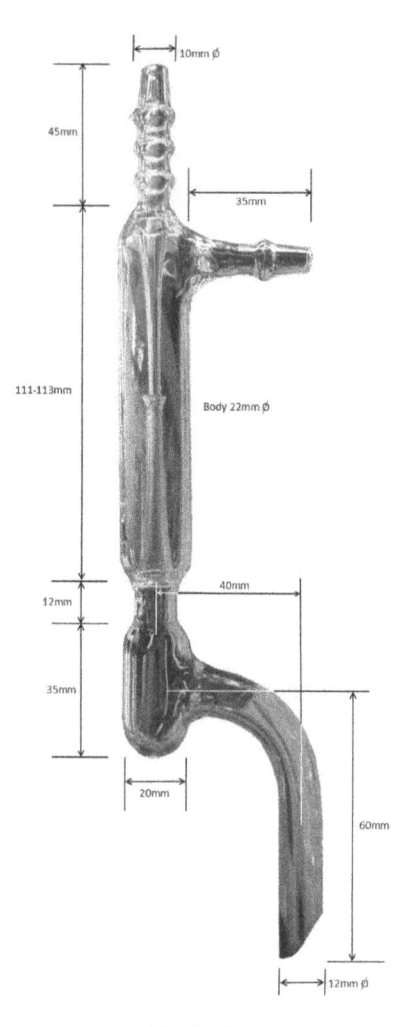

Fig. 5.161.

Button Seal

This exercise will utilise all the sections covered so far. The exercise is based upon a "water jet pump" where two jets act as a venturi to create a vacuum.

From the diagram it will be seen that the two back to back button seals will have to be central to the body and the two tapered jets made to a specific length and diameter (Fig. 5.161). The side arms and jets will be made separately and once the body has been prepared assembly can take place. Part of the main jet will be made using a V block to reduce the diameter of a tube

Fig. 5.162.

and to form a shoulder. V blocks are available commercially and are quite often free standing. I prefer to have one screwed to the side of my burner so that it is close at hand. It will be found to be quite useful not only for this exercise but also for shaping, reducing and occasionally for a short period — resting the glass midwork sequence.

It is advisable to first practice using the V block: take a 30 cm length of 20 mm ø tube. Mark the centre and rotate the glass in the top of a gas, air, oxygen flame, as the glass starts to soften it will start to constrict, withdraw from the flame and rotate constantly as you lightly rest the left side of the constriction on the V block and pull slightly to initiate the formation of what will be a shoulder (Fig. 5.162). Reduce the size and intensity of the flame and aim at the point where the V block had made its mark, as the glass softens once more withdraw and repeat restricting the diameter of the glass, remembering to rotate evenly and not to force the glass against the V block. The technique requires a light touch and an awareness of just how much one wants to constrict the glass. In this case down to 12 mm ø with a rounded shoulder to the left of the V block (Fig. 5.163). The constricted tube to the right of the shoulder can be reworked to regain an even parallel tube. As we only require a short piece of tube coming from the shoulder the technique alleviates the need to make a separate unequal join. A notable feature of the pump are the "olives" on the side arm and on the water inlet which enables rubber or plastic tubing to grip the glass.

Fig. 5.163.

First Olive Exercise

Take a 30 cm length of 8 mm ø tubing, use a well aerated gas, oxygen flame and position the glass at the very top of the flame, concentrate on your thumbs as you rotate the glass using the left hand over and the right hand underneath grip until the glass is felt to soften at which point start to push together with each rotation to form a button while the glass is still at the top of the flame (Fig. 5.164). Withdraw from the flame and put a little more air in the flame, then position the flame so that the left edge of the flame is level with the right edge of the button, continue to rotate the glass and as soon as it is felt to move withdraw immediately, pause, then pull gently as you rotate.

The glass will feel stiff and there will be a resistance for it to move, but on inspection there should be a taper leading from the base of the button (Fig. 5.165). Practice until confident of making a clearly defined button and taper. The taper will be cut at its narrowest point and glazed.

Second Olive Exercise

The side arms are made two at a time from 8 mm ø tubing 28 cm long, mark the centre and push up a button 30 mm to the right of the mark, then a taper close to the right of the button. Place to one side and repeat on a further two

Fig. 5.164.

Fig. 5.165.

lengths of tube, then return to the first which should now be cool, with the premade button in the left hand push a second button 30 mm to the right of the centre mark, then the taper to the right of the button (Fig. 5.166).

Allow to cool, scribe a mark at the centre and snap cut to make two side arms complete with olives and handles (Fig. 5.167), repeat with the 2nd and

Fig. 5.166.

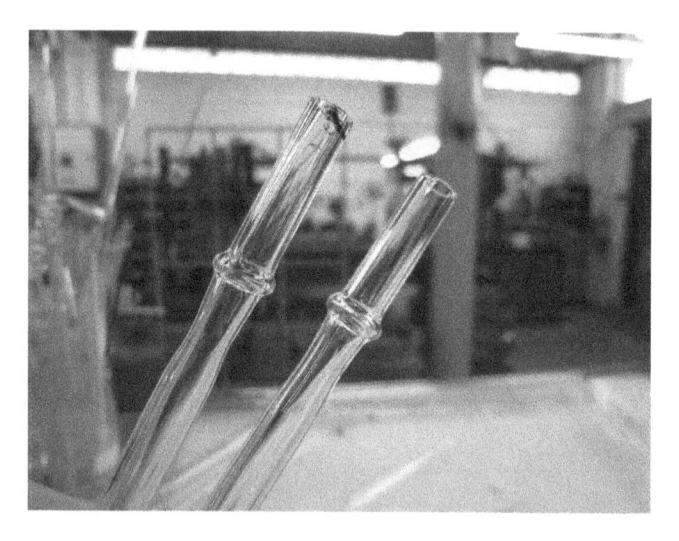

Fig. 5.167.

3rd tubes. The side arms will be attached in the normal manner and also by the "hot one melt" method. Take one of the pre-prepared side arms with the longer length in the left hand, using either a tapered faceted carbon or brass reamer lubricated with beeswax. Ream out a taper from the edge of the open tube to the button, then thicken the edge by heating and rotating against the flat side of the reamer or a piece of flat carbon plate to create a definite thickness at the mouth tapering back to the button (Fig. 5.168).

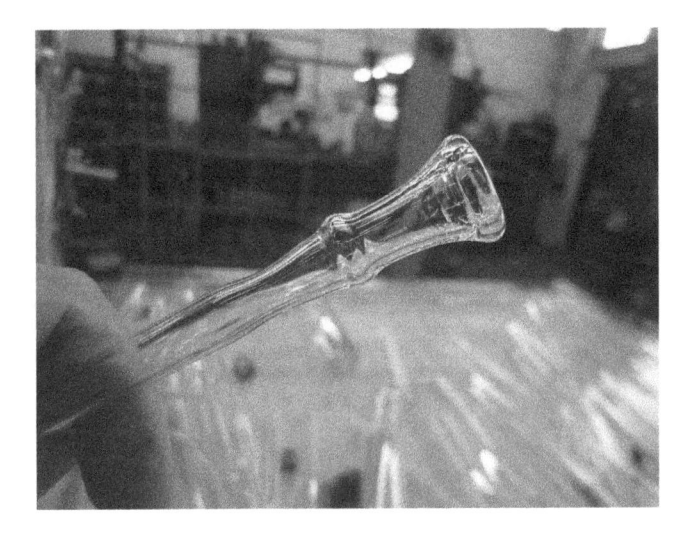

Fig. 5.168.

Third Exercise — One Melt

Using the prepared sidearm we will practise the "one melt" side arm. Take a piece of 20 mm ø tube, seal one end and blow or pull a hole slightly smaller than the prepared end of the side arm. Seal both ends of the 20 mm ø tube and position the hole behind a hot gas, oxygen flame with the side arm in front of the flame, flash the flame across the opening and watch as it starts to melt back and widening the hole meanwhile the edge of the side arm has been rotating in the flame which should by now be very hot, move both pieces a fraction to the left of the flame and push the end of the hot side arm onto the equally hot hole (Fig. 5.169).

Bring the glass towards you and blow to regain the diameter. Continue to hold the side arm in position as it will still be hot at this point giving time to correct any misalignment. This form of hot side arm can be quite satisfying to make as one is using the very characteristics of the hot glass to flow into the join which is shaped by surface tension (Fig. 5.170).

Fourth Exercise — Water Inlet

Starting with 30 cm length of 10 cm ø tubing, mark the centre and push up a button, then a second and third adjacent to each other, finally make a taper

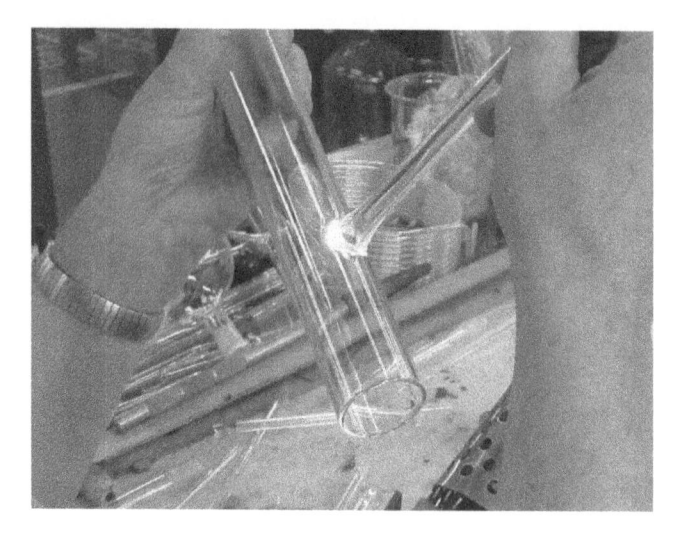

140 *Laboratory Scientific Glassblowing: A Practical Training Method*

Fig. 5.169.

Fig. 5.170.

from the third button (Fig. 5.171). Repeat the exercise two or three times, then return to the first which should now be cool enough to handle. Place the "olives" in the left hand, move 10 mm to the right of the row of olives and push a button, this will be used to make the seal, soften the flame and pull a taper (spear point) to the right of the seal button, measure 57 mm

Fig. 5.171.

Fig. 5.172.

along the taper from the button and when cool mark and snap cut. The diameter at this point should be approximately 2 mm ø (Fig. 5.172).

Fifth Exercise — Large Jet

Prepare the large jet from 20 mm ø tubing by first pulling a spear point, then second leaving 60 mm between each shoulder, using the V block constrict the

Fig. 5.173.

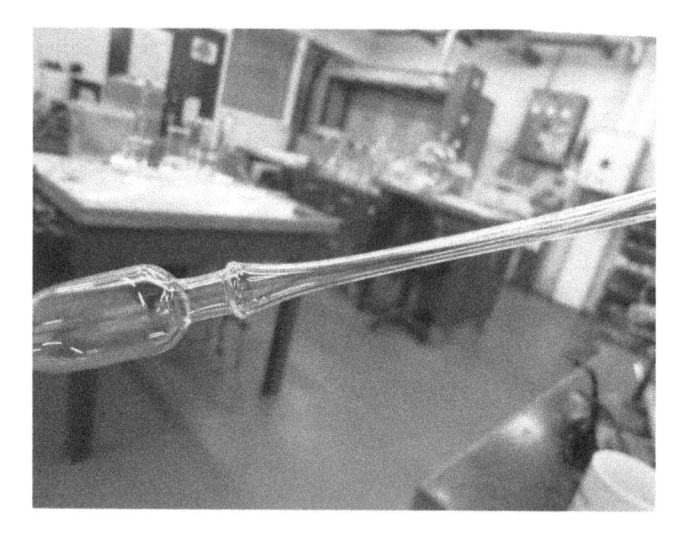

Fig. 5.174.

tube at the midpoint and form a shoulder leaving 30 mm of straight tube to the left (Fig. 5.173), make sure that there is 15 mm of straight tube to the right of the shoulder which is 12 mm in diameter. At this point push up a button, then move to the right where the remaining larger tubing is and pull a spear point from the larger diameter (Fig. 5.174) to make a slower taper than the

Fig. 5.175.

first jet. Measure 57 mm along the jet from the button, mark and snap cut at this point, then using a fine carbon or brass reamer form a small flange or funnel. This is a delicate operation due to the small diameters involved. Therefore, heat only the very end of the cut tube (Fig. 5.175). All this assumes everything is inline; one may as an alternative method use a piece of 12 mm ø tube to push a button, pull a taper, cut to length and prep the small funnel.

Assembly — Body

Take a piece of 22 mm ø tube 30 cm long, mark the centre, pull a spear point, burn through the centre of the point, allow to cool, take one body, heat the point at the shoulder and burn off, reheat and blow to form a rounded end, reduce the size of the flame, heat the centre of the domed end and blow out (Fig. 5.176). Insert the large jet inside the prepared hole making sure that the button sits on the hole and is not too proud or the hole too large (Fig. 5.177), heat the seal area and as the glass is felt to soften push together gently, then run the seal in to melt the two pieces together, withdraw and blow while all the time imagining it to be just a simple unequal join.

Return to the flame, rotate, melt and incline the left hand, this will encourage the glass to drop into the shoulder. Blow to ensure a rounded shoulder

144 *Laboratory Scientific Glassblowing: A Practical Training Method*

Fig. 5.176.

Fig. 5.177.

which comes off the seal at right angles (Fig. 5.178). As the glass cools, rotate and watch the inner jet and centralise it if necessary by slowing or stopping to allow the jet to sag into the centre. The seal can be warmed and the jet centralised once more until eventually successfully aligned. Measure the body and at 11 cm from the shoulder, pull a spear point and make a rounded end at this

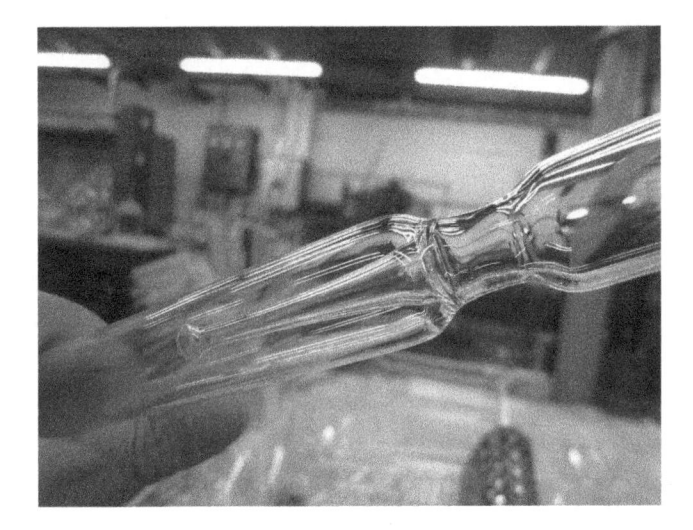

Fig. 5.178.

Fig. 5.179.

point (Fig. 5.179). Blow a hole and insert the smaller jet, check that the tip of the jets will fit inside each other; if all is well make the second seal. On completion the smaller jet should not only be central but also have an even annular space (Fig. 5.180). The tip of the jet should just be inside the funnel and not touching (Fig. 5.181). While the seal is still hot blow a hole in the body close

146 *Laboratory Scientific Glassblowing: A Practical Training Method*

Fig. 5.180.

Fig. 5.181.

to the shoulder and attach the side arm using either method, flame anneal and allow to cool (Fig. 5.182). Return to the 20 mm ø tube at the bottom of the pump and make a rounded end, take a 15 cm length of 12 mm ø tubing to make the water outlet. Blow a hole in line with the first side arm and attach the

Fig. 5.182.

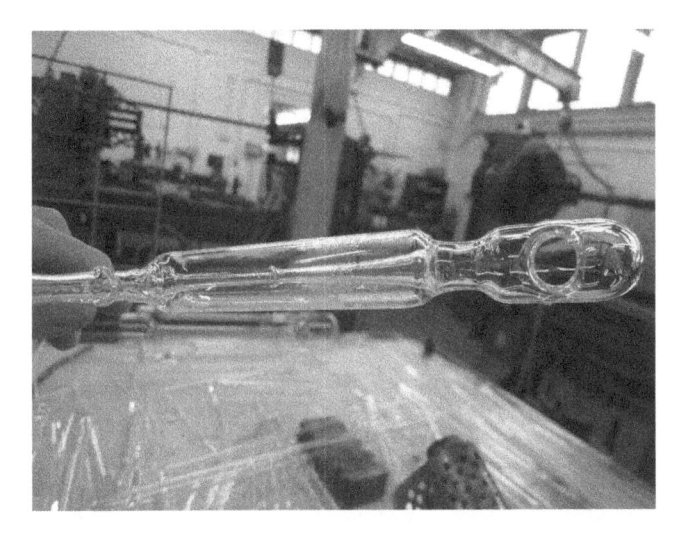

Fig. 5.183.

12 mm ø tube, once run in make a right angle bend, check once more the alignment with the first side arm (Fig. 5.183), allow to cool, measure and mark the side arm and cut to length by initiating a crack with a hot point (Fig. 5.184), then snap cutting. Cut the water inlet and first side arm to length and glaze all three cut surfaces (Fig. 5.185).

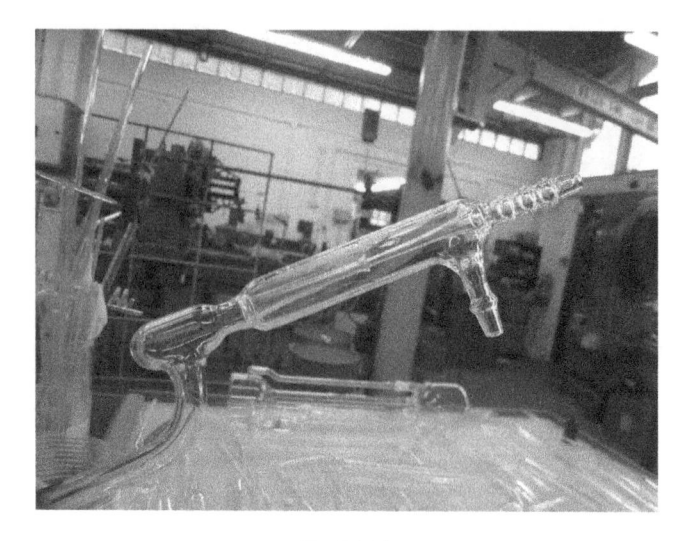

Fig. 5.184.

Fig. 5.185.

Repeat the whole exercise once more, then start with the smaller jet and side arm finishing with the large jet. Reversing the sequence will make you think — especially which end to seal prior to blowing.

5.12 Bulb Blowing

One only has to think of a soap bubble to realise due to surface tension the most natural shape for molten glass to form when blown in a sphere. There are many ways of producing spheres, which are referred to as "Bulbs".

To begin the exercise select the red rubber bung with the 9 mm ø tube through the centre we have previously made in Section 5.3. We will use this to simulate the weight and rotational feel of the bulb that we are about to make.

Using the left hand rotate the glass and bung using the thumb and forefinger, raise the hand a little and put pressure on the wrist to maintain the 90° wrist/forearm. We are about to practise a new type of rotational control keeping the same hand position, roll the tube down the length of the forefinger on reaching the tip rotate the tube in the opposite direction, back along the forefinger and onto the inside of the thumb on almost reaching the tip of the thumb reverse the action, repeat back and forth along the inside of the thumb and forefinger. Mark a line along the length of the bung, resume rotating the glass and bung back and forth while observing that the bung is rotating two or more times in either direction. One may find that this form of rotating the glass is easier and gives greater control.

"Do Not" now abandon the primary method as one will in time be able to combine the two styles together. Using either technique pull a series of spear points from 15 mm ø tubing leaving 40 mm between each shoulder. Open one end and place in the right hand — ensure that the spear points are running on a true central axis. Using a flame 20 mm wide heat a 20 mm length on the right side of the main tube until molten, do not allow the hands to wander apart, preferably keep a slight pressure together, withdraw from the flame and rotate using either method while the glass cools a little and the colour darkens — drop the left hand, hold the glass vertically and blow hard against the rapidly cooling glass to produce a slight increase in diameter and a sharp shoulder (Fig. 5.186), centralise the spear point in the left hand and seal the right. Allow to cool before transferring the newly sealed point into the left hand, heat the right side of the body, withdraw, rotate, pause, blow to increase the tubing diameter similar to the tubing to the left (Fig. 5.187). Return to the flame and aim at the centre noting that the edges of the flame will be playing on the still warm tube. Soften the flame with a little air and using the top of the flame until the glass softens withdraw, pause and blow

Fig. 5.186.

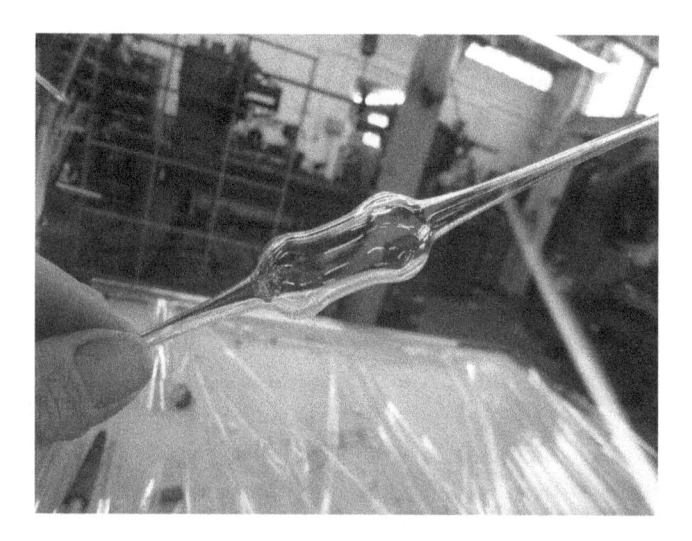

Fig. 5.187.

with the glass held vertically (Fig. 5.188), watch as the diameter increases, return to the horizontal. Play the flame on the right side of the preformed sphere — heat only one hemisphere (right-hand side only) withdraw/rotate/blow to form a hemisphere which will (with practice) be evenly formed as we have used the opposite shape to support the other.

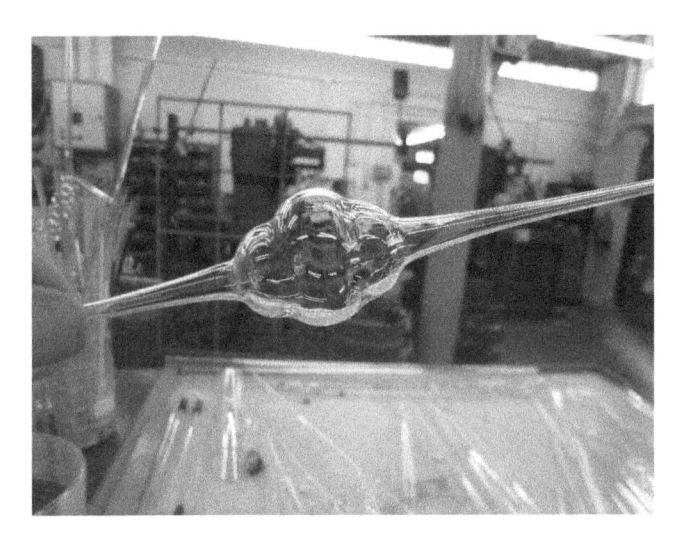

Fig. 5.188.

Open the left hand point and seal the right, allow to cool. Transfer the sealed spear point into the left hand and repeat the operation on the right-hand side of the sphere (Fig. 5.189). This somewhat tedious method will produce a bulb of even shape and wall thickness. One would have felt the momentum as the bulb is rotated on the spear points. One may also note that there is an increase in radiated heat off the spherical shape. One's hands must and will get used to the heat. One good side effect is that it will encourage the pulling of longer and stronger spear points which will enable the hands to be positioned a little further apart.

This exercise will improve the control and confidence while developing an understanding of the material benefiting the students' ability to tackle the following exercises.

The "splash head" exercise now combines four previous exercises.

1 Unequal join
2 Bulb blowing
3 Button seal
4 Bends

Tubing required 30 mm ø for the "body" and 8 mm ø for the seal and bends. Pull two straight spear points from the 30 mm ø tubing leaving a

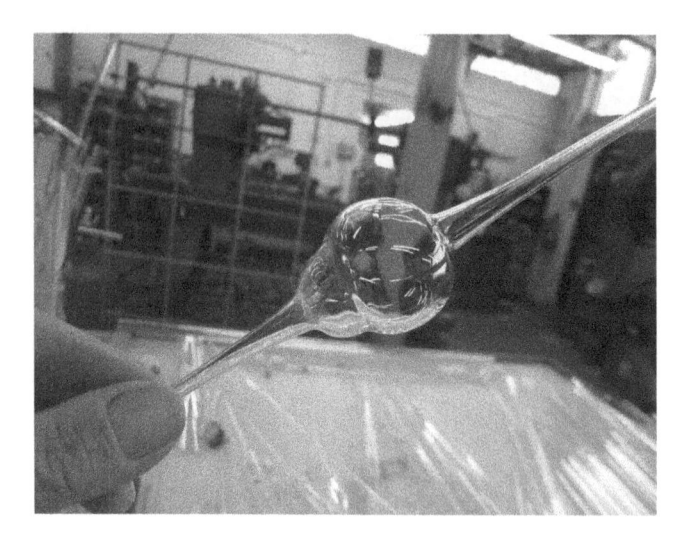

Fig. 5.189.

straight portion 30 mm between each shoulder, straighten the spear points to ensure they are lying along the true central axis. Choose the longer, stronger and straighter point as the handle. Open the tip of this spear and remove the opposite spear, blow a hole and proceed to make an unequal join with a 120 mm length of the 8 mm ø tubing (Fig. 5.190). Once the initial join has been "run in" increase the size of the flame to 25 mm wide and proceed to blow both the shoulder and body into a hemispherical end by controlling the right hand edge of the flame so that the 8 mm ø tube is not overmelted as the bulb is being formed from the body and shoulder. Correct for central alignment as the glass cools by rotating. Transfer the 8 mm ø tube to the left hand and seal and open the end of the spear point. Proceed to thicken and constrict the shoulder of the spear by reducing the flame width a little and rotating evenly — withdraw from the flame and centralise as the glass cools once more. Return the glass to the flame which should be returned to 30 mm wide. Soften the flame by reducing the amount of air/oxygen and heat the whole sphere while concentrating on the centre — do not allow the hands to drift apart as a very slight pressure together is best as one feels the glass softening — slow the rotating a little and combine with a back and forth motion as the tube rolls along the finger and thumb — left hand dominant with the right hand following in time with each movement. Withdraw and

Fig. 5.190.

blow as the glass cools, a bulb diameter between 40 mm and 45 mm is quite sufficient at this point (Fig. 5.191).

Once satisfied with the bulb remove the spear, pull away any excess glass and blow a hole 9–10 mm diameter (Fig. 5.192). Warm off using a feather flame and put to one side as we prepare the seal and inner bend. Take a piece of 8 mm ø tubing 240 mm long, mark the centre and push a button using the very top of a flame 5 mm wide, then seal the end of the tube to the left. Increase the width of the flame to 30 mm and soften with air, make a right angle bend immediately to the right of the button, i.e. warm the glass, feel it soften, cork to the ceiling, make the bend, blow — allow to cool, using the glass knife scribe a mark on the inside of the bend 15 mm to the right of the vertical (Fig. 5.193).

One has to imagine where the end of the bend will be after the button seal is in place and the bend within the sphere! Snap cut the glass and flame polish the end. Return the sphere to the left hand with the 8 mm tube sealed. Insert the bend into the bulb and rest the button on the opening (Fig. 5.194). When all is well proceed to run in the seal using a flame 5–10 mm wide. Concentrate on the seal and the tube to the right — once one can see that the seal is made withdraw and blow while holding the glass vertically, inspect the seal to ensure that it is showing an even dark line at the point of contact.

Fig. 5.191.

Fig. 5.192.

Reshape the sphere to ensure the seal is sitting in line with the arc of the sphere and not proud. Soften the flame and increase to 20–30 mm wide and proceed to make a bend immediately to the right of the seal. As the glass softens withdraw from the flame — hold the glass vertically with the left

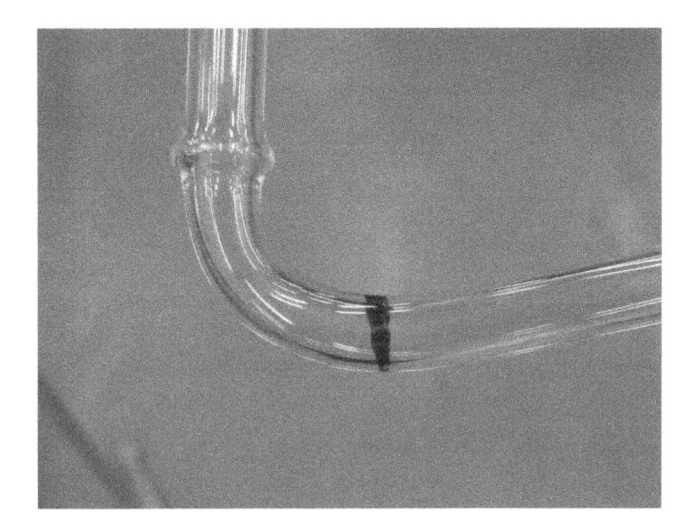

Fig. 5.193.

Fig. 5.194.

hand ensure that the inner bend is facing away and make the bend with the right hand in line with the "inner" and blow (Fig. 5.195).

View the splash head from the side and depress the tubing in the right hand below the 90° angle — in time the bend and angle will be formed in

Fig. 5.195.

one swift action. Flame polish the ends of both lengths of the 8 mm ø tube (Fig. 5.196).

During the exercise we have constantly changed the position of the glass to ensure that the weight has been in the left hand so as to establish a standard method of working whatever the configuration of the glass. Also as the exercise has progressed the need to control size, shape and intensity of the flame has become more apparent.

Pear-Shaped Column

The column consists of a series of pear-shaped bulbs with an inverted ridge which when in use collects the condensing distillate. The number of bulbs can vary from 4 to 12. The exercise will consist of four bulbs.

The bulbs will be blown directly onto 15 mm ø tubing with an 8 mm distillate side arm. We begin with a 15 mm diameter tube 200 mm in length. Mark the centre point, seal one end and place in the left hand. A flame 25–30 mm wide is softened with air and oxygen is used to make a bulb "in one go". Aim the flame at the centre mark, hold the glass near to the top of the flame, as the glass begins to soften push together slightly to form a

Fig. 5.196.

Fig. 5.197.

"bulge" (Fig 5.197). Concentrate on rotating evenly and constantly — feel the glass soften, keep the glass in the flame past the point at which you want to withdraw, drop the left hand and blow to produce a bulb approximately 30 mm in diameter (Fig. 5.198), keep the hands together as you blow, return

158 *Laboratory Scientific Glassblowing: A Practical Training Method*

Fig. 5.198.

to the horizontal to inspect for central alignment, rotate to centralise as the glass cools.

Once satisfied reduce the size of the flame to 5–10 mm wide, aim the flame on the right-hand side of the bulb at the point where it joins the 15 mm ø tube but not at the 15 mm tube which must remain solid, as the glass softens push together, watch the bulb indent, withdraw and continue to rotate and push together as the glass solidifies (Fig. 5.199).

Soften the flame and rotate the left-hand side of the bulb in the top of the flame, when the glass starts to soften gently pull apart while rotating to form the taper. Rotate to align. The exercise to make a single pear-shaped bulb is complete (Fig. 5.200).

Repeat the exercise starting with 15 mm ø tube cut to a length of 270 mm, place a cork in one end and make 150 mm from this point, place the sealed end in the left hand, blow the bulb, make the indent and taper. Imagine a point 30 mm to the right of the first bulb, proceed to complete a second bulb at this point with indent and taper, allow to cool (Fig. 5.201). A third bulb is made in the same way from the left towards the right while starting from a tube 330 mm long. Finally a fourth bulb column from a piece of 15 mm ø tube 400 mm long. The initial bulb becomes more and more difficult as the right hand has to support and control the longer tubing.

Fig. 5.199.

Fig. 5.200.

Gradually increasing the number of bulbs from one to four will be challenging but rewarding (Fig. 5.202).

The top tube set at 110° from the vertical column, the top may have a rim made with a carbon or reamer (Fig. 5.203). In the first sentence of this section I referred to the blowing of a soap bubble. Not only is a sphere a

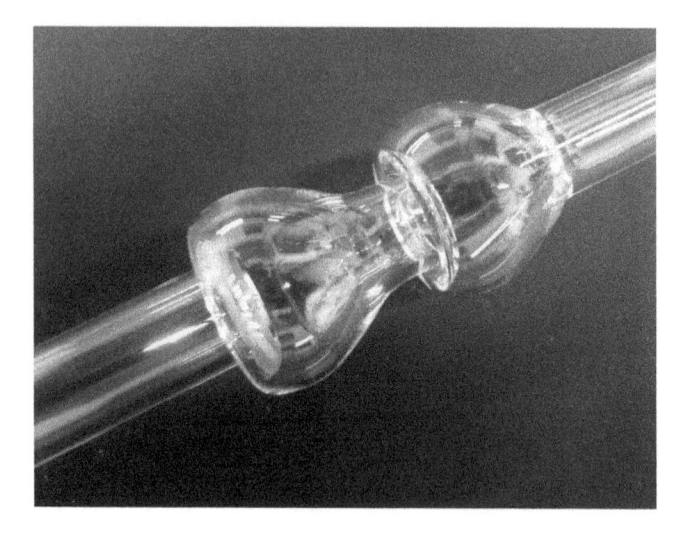

Fig. 5.201.

Fig. 5.202.

natural shape formed by molten glass. The "controlled" blow of a soap bubble will create the larger bubble. Likewise control of the blow either duration or force will have a direct effect upon the size, shape and thickness of a glass sphere.

Fig. 5.203.

5.13 Spiral Winding

Method 1

Use of a metal mandrel (former)

Mandrels are available commercially and consist of a steel tube of various lengths and diameters attached to a heat-resistant handle — often wood. They can be made simply given the correct engineering facilities which can be quite helpful especially when specific diameters are called for.

One will find either a drilled hole or peg close to the handle to which the glass is attached. The mandrel surface should be smooth and not distorted. The softened glass tubing or rod will be wound around the mandrel. To support the weight of both the glass and mandrel a simple V block or double wheel support will be used. Positioning of the tubing in relation to the flame is critical as the intention is to soften the glass "just" sufficiently for it to wrap around the mandrel without distorting the tubing diameter. There is a fine balance as too hot a flame will cause a flattening of the tube, if too cold the tubing will be unable to wrap around the mandrel and will split and break.

We will start with a mandrel of some 25 mm diameter. The glass tube will be a full 1.5 m length and 8 mm ø. The end of the tube is prepared by heating at a point of 40 mm from the end with a soft flame, once the glass is seen to "sag" a gentle bend is made by gripping the end of the tube with tweezers and twisting to just past 90°, the bend will be used to hook onto the mandrel and will be removed on completion of the spiral. Place the tube to one side while adjustments to the flame are made. The support is placed to the right of the flame; the mandrel handle will be to the left. The flame should be feather-like and not a hot laminar flow (pure oxygen/gas) flame, neither should it be a soft luminous cool flame, ideally a transparent gas/oxygen/air "noisy flame", if no compressed air is available an overgassed (little oxygen) flame should suffice. Adjust the angle of the burner so that the feather-edged flame is almost horizontal, the glass will be fed along the top of the flame length and the bottom edge of the mandrel will also be positioned just above the flame (Fig. 5.204).

The mandrel must be prewarmed to expand the metal prior to winding the spiral, once the mandrel is warm, hook the 8 mm tubing through the hole in the mandrel. A standing position is advisable, hold the handle in the left hand while the mandrel rests upon the support, the right hand lightly supporting the tube as it lies along the top of the flame. One will both see and feel the glass soften, start to rotate the mandrel so that the softening tube is pulled gently onto the mandrel from underneath, continue to rotate the mandrel while feeling a slight resistance as the glass softens, after the first

Fig. 5.204.

Fig. 5.205.

rotation one can gauge the distance between each spiral — which is called the "pitch" (Fig. 5.205).

To gauge the angle of the pitch and the distance between each turn of the spiral — look through the flame and concentrate on the bottom of the mandrel at the point where the glass disappears underneath. The angle between the adjacent spiral and the tube will be observed.

Control the evenness of the pitch by slightly raising and lowering the mandrel from the horizontal. Guidance with the right hand as the glass passes through will also control the distance between each spiral. One will have to move the whole mandrel gradually to the left as the spiral length increases (Fig. 5.206). To separate the spiral from the mandrel allow the work to cool, this will enable the metal mandrel to "shrink" away from the glass. Scribe a mark with the glass knife where the glass enters the hole in the mandrel and tap the mark with the back of the knife to break the tube in a controlled manner. The glass should at this point slide off the mandrel, if not allow to cool further.

If the glass remains on the mandrel it would suggest that the mandrel was positioned too deep in the flame rather than above the flame and that the glass was also too hot. If the spiral remains stubbornly attached one will have to resort to breaking the entire spiral off the mandrel which in turn will require attention to remove any glass still adhering.

Fig. 5.206.

In making the spiral there is a four way balance that has to be taken into consideration. They are as follows:

- The speed of rotation
- The diameter and wall thickness of the glass
- The intensity of the flame
- The position of the mandrel and the glass in relation to the flame

Free Hand Spirals

Flat spiral

As the name suggests spirals which are parallel, tapered or flat can be made with practice without the aid of a mandrel. One would have seen in the previous exercise that no blowing was involved, likewise spirals made from smaller diameter tubing blowing should not be necessary.

To begin we will use rod rather than tube to enable the student to concentrate upon flame position and the angle at which each successive turn of the spiral is aligned one to another. Take a standard length of 6 mm rod and cut in half, measure 250 mm from one end and make a mark, with the longer length in the left hand heat the mark at the top of a "noisy" gas, air, oxygen flame until one can feel the glass soften, withdraw and make

Fig. 5.207.

a 90° bend and allow to cool. Transfer the shorter length into the left hand and soften the flame further with air, hold the glass in the left hand at right angles to the flame. The right hand holds the longer length of glass at an inclined angle. Only the last 30–40 mm of glass closest to the bend is heated, as soon as it is felt to soften withdraw from the flame. Rotate anti-clockwise with the fingers of the left hand to make quite a small tight bend (Fig. 5.207), as the glass solidifies drop the left hand "look" at the piece as after each section of glass is heated and bent we will have to look at the spiral from the edge and "end on" to check alignment (Fig. 5.208). The initial bend is difficult as there is little or nothing on which to align although as we progress the spiral this will become easier. Return the glass to the top of the flame, heat only 30 mm of the straight section closest to the bend and not the bend itself.

Very little time should be spent heating the glass as it only needs to be just mobile; too much heat and too soft a glass at this point will make the glass uncontrollable. Withdraw and start bending the second section while holding the spiral "end on" and gradually bend the glass so that the spiral begins to radiate outwards, return to the flame once again, heat and withdraw from the flame, as the glass is "felt" to soften, hold "end on" as you make a gradual bend on the cooling glass while making a progressively larger spiral (Fig. 5.209). Work in a relaxed manner, do not try to "force" the bend — always concentrate the flame onto the straight section and avoid returning to the bend if possible.

Fig. 5.208.

Fig. 5.209.

Parallel coil

Once more using a 6 mm ø rod prepare the rod by making the right angle bend and initial tight bend (Fig. 5.210). Using the top of a similar flame heat a 30 mm section close to the first tight bend, when the glass starts to soften, withdraw and hold horizontally while turning anti-clockwise with the left

Fig. 5.210.

hand, as the bend is made incline the right hand a little so as to induce a slight pitch in the first bend. Reheat the next section and make a further inclined pitch as the bend moves from left to right.

Confused?

It may well help if a premade mandrel formed spiral is positioned on the bench so as to aid imagination, both the pitch and direction of the bend. Inevitably the first attempts and the first part of a handmade spiral will be uneven but once one becomes used to the flame position and the amount of heat required, practice definitely will yield results.

Further coil winding will be described in Section 5.18 using a slow running attachment to a glassblowing lathe in combination with a hand torch where the glass will be wrapped over the top of the mandrel.

Not only will coils be used in condensers, but by their very shape they can be used to give flexibility to a glass system where there may be a large temperature gradient between the two static pieces.

5.14 Liebig Condenser

This condenser is named after the Nineteenth Century Chemist Justus Von Liebig who amongst others advocated the use of a cooling jacket fitted around an air condenser to improve efficiency by providing a constant flow of cooling water around the annular space. The approach to making the condenser is based upon the "supported internal seal" as in Section 5.10.

The outer jacket will be 26 mm ø
The inner 13 mm ø
The inner tube extensions 13 mm ø
The side arms 9 mm ø

Take a piece of 26 mm ø tube 660 mm long, mark the centre and pull a spear point Section 5.2 to separate into two 320 mm lengths. Take one length, remove the spear point at the shoulder and make a rounded end and place to one side. Prepare the inner by cutting a 220 mm length of the 13 mm ø tube, also cut four lengths of the 13 mm ø tube, 140 mm long for the extension tubes. Take a piece of corrugated card 300 mm × 160 mm, flatten the corrugations with the back of the glass knife and recut the card to 265 mm × 125 mm. Support the 13 mm ø inner tube using the corrugated card within the outer tube with the inner resting against the inside of the dome.

The card should just protrude past the opposite end by 10 mm which will enable it to be withdrawn on completion of the seal.

The side arms are made from the 9 mm ø tube by marking the centre point and also 30 mm on either side. Using the very tips of a gas, oxygen, air flame 5 mm wide aim the flame at the right hand mark, rotate in the flame, as the glass is felt to start softening push together while rotating and withdraw push and rotate as a small "button" is seen to appear (Fig. 5.211), increase the width of the flame to 10 mm wide by just increasing the gas, immediately return to the flame to the right of the button as the glass is felt to soften withdrawn, rotate and as the glass cools pull slightly to form a slight constriction (Fig. 5.212).

Allow to cool thoroughly, transfer the work into the left hand and repeat the button/constriction at the mark to the right of the centre line and allow to cool. Using the glass knife mark the centre between the two buttons and snap cut (Fig. 5.213), put to one side.

To make the next section we will require the previously made 26 mm ø domed tube with the supported inner 13 mm ø tube, a piece of the 13 mm ø and one of the side arms. Finally corks to fit both the 26 and 13 mm tubes. Warm the domed end in the very top of a gas/air flame, watch for the first signs of the orange sodium flame coming off the glass which will indicate that the glass is about to soften. Reduce the size of the flame and increase its intensity with the introduction of oxygen. Aim the flame at the top of the dome, then incline the glass slightly and also position the glass so as to make it easier to see the domed end melt back onto the inner tube. The dome will flatten as it touches the inner. Watch for the contact line (Fig. 5.214), once it can be

Hand Techniques **169**

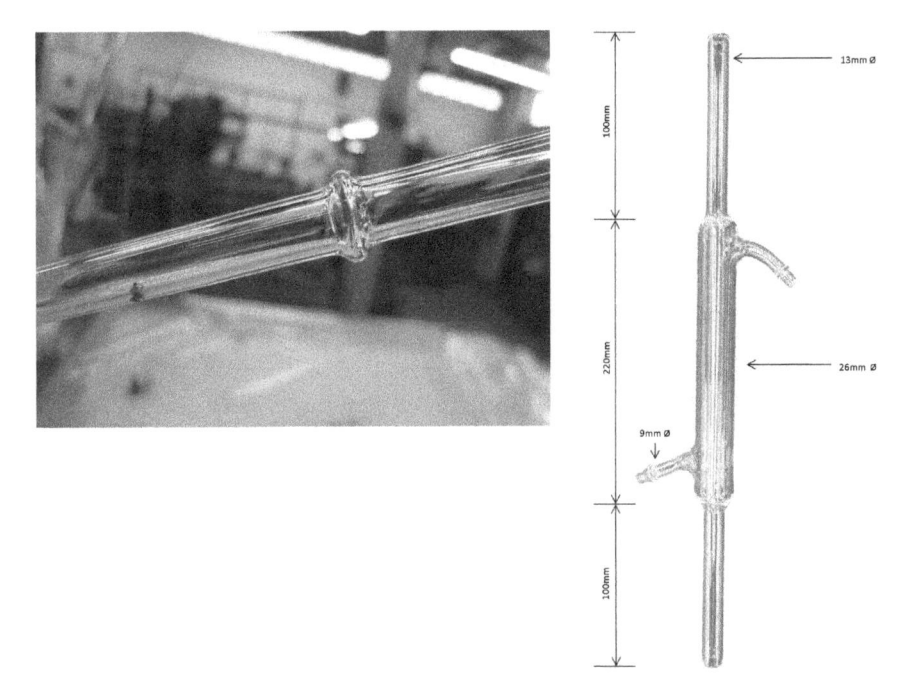

Fig. 5.211.

Fig. 5.212.

Fig. 5.213.

Fig. 5.214.

seen to be forming an even line remove from the flame and blow to confirm the seal is made, return to the flame and heat the centre once again, withdraw and "blow out" break off the thin bubble produced with the back of the glass knife (Fig. 5.215).

Place the cork in the end of the 26 mm ø tube and place in the left hand and heat the hole, while taking a piece of the 13 mm ø tube which is heated on the opposite side of the flame, when both edges are hot withdraw from

Fig. 5.215.

the flame and touch together, then a slight pull and a blow to confirm the seal is made. Reduce the size of the flame to 5 mm wide and heat the 13 mm ø tube just to the right of the seal with the left edge of the flame playing on the seal. Withdraw, rotate, pause and blow.

Do not allow the hands to wander apart. Keep the hands together, repeat a second time to make sure that the seal is well "run in" (Fig. 5.216). Increase the flame width to 10 mm and aim the flame on the shoulder with the right side of the flame on the seal. When hot withdraw and drop the left hand and blow, this should redefine the shape of the shoulder/seal, rotate to centralise. Reduce the flame to 5 mm wide, aim at a point just to the left of the shoulder, withdraw and blow to form a dome, then aim the flame at the top of the dome, withdraw and blow out to create a hole, place a cork in the end of the 13 mm ø tube. Take a prepared side arm, heat the edge and the hole, withdraw and attach to the body, in effect making a T piece join (Section 5.5). Run the join in using either the sectional method or the back and forth rotational heating method or a combination of both methods which will be both enjoyable and terrifying at the thought that you may fail at this point after so much previous work (Fig. 5.217).

Once satisfied the side arm is run in and the initial ridge has been removed, gently bend the side arm downward towards the 13 mm tube. Warm the seal/sidearm area in a feather-like flame to remove any extra strain and put aside to cool. Remove both corks and the support material — reverse

Fig. 5.216.

Fig. 5.217.

the cooled glass. Place the 13 mm ø tube in the left hand, using a 10 mm wide flame heat the 26 mm ø tube just to the right of the inner tube and pull a spear point — burn through the point 100–150 mm from the shoulder, allow to cool, then using the point as a handle heat the tube in line with the end of the inner tube and pull the 26 mm ø tube onto the inner tube until contact is made with the inner tube, reduce the width of the flame to 5 mm and pull the spear away and burn off. Blow down the 13 mm ø tube, then immediately down the sidearm to confirm the seal is made, heat the whole end once more and blow down the centre and the sidearm to form the shoulder.

Heat the end of the centre tube and blow out the end through the 13 mm tube only, then after removing the thin bubble seal the 13 mm tube with a cork, take the 2nd piece of the 13 mm ø tube and heat both the ends of the tube and the hole in a flame 20 mm wide until softened, withdraw, touch together, slight pull and blow down the 13 mm tube in the right hand and the sidearm, run in the seal with a 5 mm wide flame blowing occasionally down the sidearm and 13 mm tube. Once satisfied with the seal and join blow a bulge near to the shoulder on the opposite side to the first sidearm, then heat the top of the bulge, withdraw, blow and break off the thin bubble to make an 8 mm hole. Seal the first sidearm with tack. Take a prepared sidearm, heat both edges — touch together with a slight pull and blow to confirm the seal — proceed to "run in" the sidearm using either method, once satisfied the sidearm is run in, soften the flame and heat the sidearm with a back and forth movement, when the glass softens withdraw and hold the condenser body at 40° with the sidearm uppermost make a gradual bend and blow (Fig. 5.218).

Allow to cool, then cut the sidearms off in the middle of the constriction and flame polish the ends. Measure 100 mm from the end of each seal, mark with the glass knife, moisten the mark and gently press the end of the tip of a heated 6 mm rod — see Section 5.1 to generate a crack which may travel all the way round if not grip the end and snap cut the tube. Flame polish the ends to complete the exercise.

The sidearm bend is interesting in that it is neither a right angle nor a "U" bend but halfway between the two, therefore the "hold" in the left hand is neither vertical as in the right angle bend nor horizontal as the start of the U bend. One will use this angled hold with the left hand more and more especially when bends have to come off close to the body, the benefit being that you can see exactly from where and to where the bend has to go, while the sweeping bend is aided by gravity as the softened glass falls into the bend.

I prefer to increase the size of the body close to the seal so as to ensure that the outer tube comes away from the seal at right angles (Fig. 5.219) which avoids any chance of the seal being very shallow and make it more vulnerable especially where temperature differences are involved between the vapours within the condenser and the cooling liquid passing through the annular space.

Fig. 5.218.

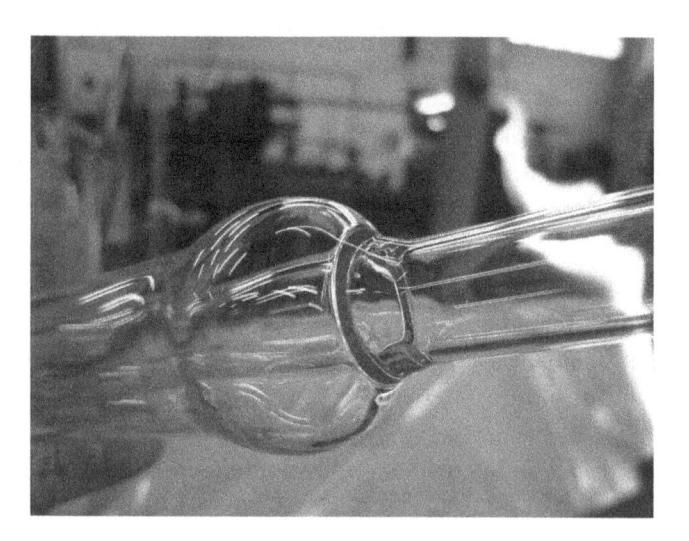

Fig. 5.219.

Not only will this shape of seal give a cleaner, stronger join, but the rounded increased diameter of the outer jacket acts like an expansion bellows.

Sweeping bends and rounded shoulders are more in tune with the liquid glass when a piece is being made and reduces the intensity of strain in glass when in use.

5.15 Spiral Condenser

Before making a spiral or coil condenser we will have to practise making a side seal which is where an internal tube is sealed through the wall of the condenser at 90°. Cut and prepare a 30 cm length of 24 mm ø tube and flame polish (glaze) each cut end. Select and cut a 28 cm length of 7 mm tube and make a right angle bend half way down its length. Prepare a piece of corrugated paper 24 cm by 16 cm. First cut an oversize piece and soften the corrugations with the back of the glass knife as described in Section 5.10. Wrap the corrugated paper around the 7 mm tube and insert within the 24 mm ø tube with the bend protruding from the top at right angles, mark the 7 mm ø tube where it aligns with the inner wall of the larger tube, withdraw the 7 mm tube and with the glass knife scribe a mark on the inside of the bend in line with the first mark, moisten and snap cut, remove any sharp points and glaze the end, replace the 7 mm tube in the packing to confirm that the bend will slide inside the 24 mm ø tube (Fig. 5.220).

Position the packing so that the inner tube is touching the inside of the larger at a point 5 cm from one end, seal the 24 mm tube with a cork after ensuring that the inner tube is projecting 6 cm from the packing (Fig. 5.221). Prepare a piece of 8 mm ø tube 12 cm long, glaze and flare one end slightly — this flared end will be used to make the side seal (Fig. 5.222).

Fig. 5.220.

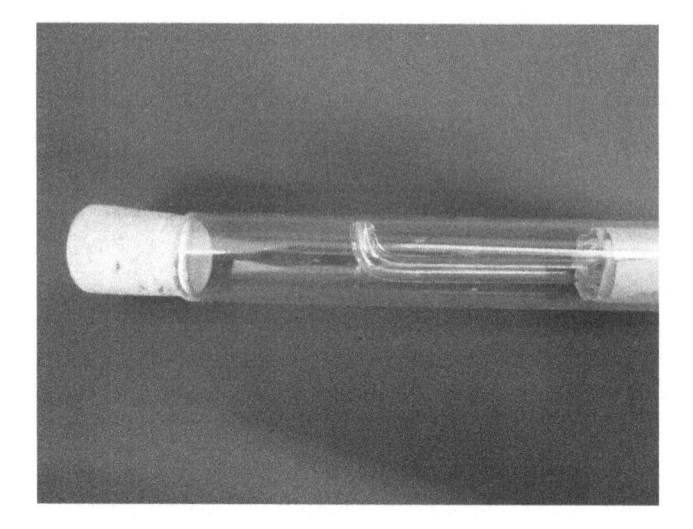

Fig. 5.221.

Fig. 5.222.

Hold the larger tube in the left hand and warm the tube in the flame at the point where the seal is about to be made until small flecks of the orange sodium flame first appear, reduce the diameter of the flame to a fine point and aim at the centre of the 7 mm ø tube, soon the outer tube will touch the inner, when it can be seen that contact has been made, withdraw from the

Fig. 5.223.

flame and blow while watching the point of contact as the outer tube regains its parallel sides which may well have sagged a little during the sealing stage — inspect the seal which should show as a dark grey ring (Fig. 5.223), if the ring is a complete circle we can move on, if incomplete return to the flame and aim once more at the point of contact.

Once satisfied take a piece of 5 mm ø rod and warm the tip slightly, then aim the flame at the centre of the seal until it just starts to glow, withdraw from the flame no more than 25 mm and lightly touch the centre of the seal with the tip of the rod and pull to make a small cone, burn the rod off and aim the flame once more at the top of the cone which should melt first rather than the surrounding tube, blow down the 24 mm ø to produce a thin bubble (Fig. 5.224) and slice this off with the glass knife.

Place a second cork in the open end of the 24 mm ø tube. Have to hand a length of carbon/graphite rod, the end of which has been ground to give a four sided tapered point to act as a reamer. Hold the 24 mm ø tube slightly to the rear of the flame so that we may see the seal hole as we aim the flame at the point where the inner and outer meet, rotate the small pointed flame around the rim of the hole and watch as it melts and flows one into the other (Fig. 5.225). Good flame control is needed as the inner tube is so much smaller and will tend to melt quicker than the outer with the consequence that

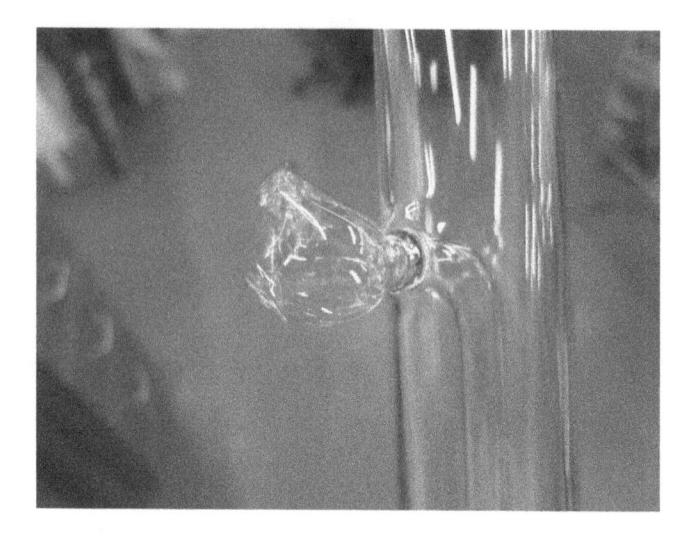

Fig. 5.224.

Fig. 5.225.

it may well collapse. Should this happen retain its inner diameter by inserting the carbon reamer while rotating between the thumb and forefinger.

Once satisfied that the seal is "run in", warm the flared end of the 8 mm ø "side arm". Once more hold the larger tube slightly to the rear of the flame with the hole touching the left side of the flame while rotating the flare of the "side arm" in the right-hand side of the flame (Fig. 5.226), when both are

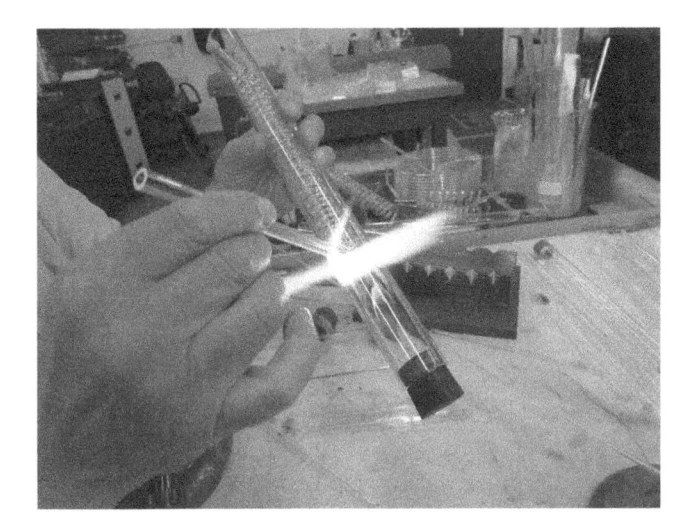

Fig. 5.226.

Fig. 5.227.

seen to glow orange withdraw from the flame and immediately "touch–pull–blow" the side arm, inspect to make sure that the seal has been made as indicated by the complete circular grey ring (Fig. 5.227), finish the side arm using either method as described in Section 5.5.

Fig. 5.228.

Practice the "side seal" a number of times to gain confidence.

The spiral or coil condenser.

Make a coil 19 cm long and 23 mm outside diameter using a mandrel 13 mm diameter and glass tubing once more of 7 mm ø. The ends of the spiral are now prepared as one end has a drip bend near the side seal, this is to encourage any distillate to run or drip from the centre of the spiral when in use. Seal with tack the end of the spiral that was attached to the mandrel, hold the spiral itself and using the unbent tube at the opposite end make a 90° bend parallel to the spiral before making a tight "U" bend (Fig. 5.228), followed by a further bend at 90° to the coil (Fig. 5.229). Cut off the excess tubing after first trying the spiral inside the outer body while allowing for the annular space between spiral and inner diameter of the body. Mark with a fibre pen the proposed length of the spiral and also an alignment mark with the first bend as both "side arms" should be in line, this will help to know in which direction to make the final bend (Fig. 5.230).

Cut the spiral to length at a point where the final seal will be made. With a second piece of 7 mm ø tube which is attached to the spiral and "run in", then make a tight right angle bend to align with the first bend and cut (Fig. 5.231) to length so that both spiral ends will fit inside the body.

There are a number of ways to support the spiral within the body, but at this stage it will be best that we continue using corrugated paper as packing.

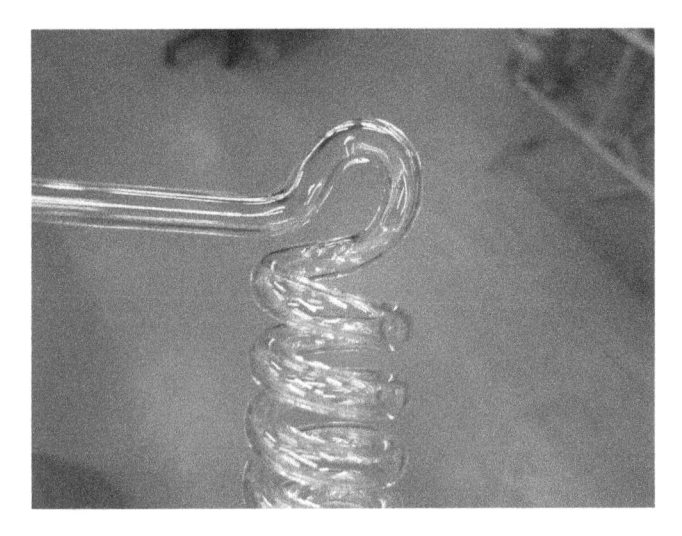

Fig. 5.229.

Fig. 5.230.

Rather than wrap the packing all the way around the spiral fold it back upon itself leaving a small gap where the side arms align so as not to interfere with the second seal or damage the first seal when the packing is being removed (Fig. 5.232). To complete the exercise we will use a "straight" body of 34 mm ø 38 cm in length.

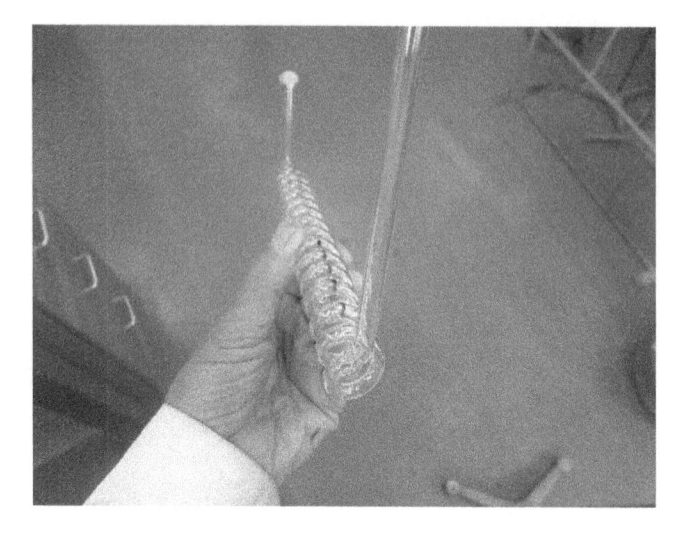

Fig. 5.231.

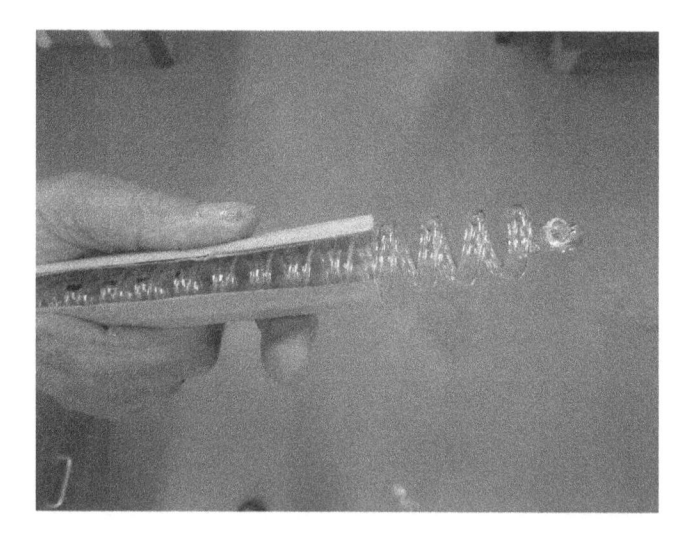

Fig. 5.232.

Prepare a number of side arms, i.e. more than two. Support the spiral within the body, seal the end with a cork near to what will be the first join. The support material should cover half the length of the spiral while overlapping the end by 2 cm which will ensure that using tweezers and gripping this excess packing removal of the support material can take place. Warm the seal

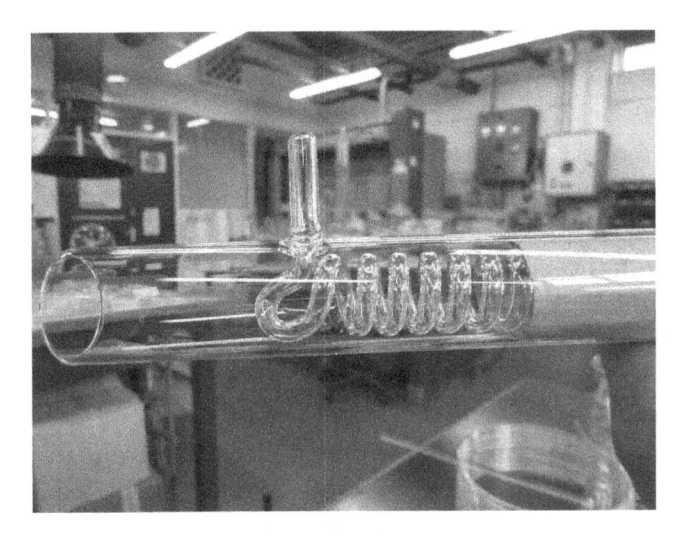

Fig. 5.233.

Fig. 5.234.

area and once the flecks of the sodium flame are seen, start to make the "side seal" and attach the "side arm" which is cut to a length of four centimetres and finally glaze the end (Fig. 5.233). Remove the cork and warm the whole seal area until the flecks of the sodium flame can be seen once more. The glass is then allowed to cool. Carefully remove the corrugated paper packing, the spiral is now supported on the first seal (Fig. 5.234).

Replace the cork in the body this time at the second seal end, warm the second seal area, then aim at the centre of the proposed seal with a fine flame until it makes contact. One must now alternate between blowing down the side arm, then the larger tube as the two entrants are now sealed and separated from the body. Once more aim at the centre of the seal and blow down the side arm only, to produce a small bubble which is sliced off with the glass knife, play the flame on the initial seal area to "run in" the join, blow down the body and occasionally use the carbon reamer to maintain the smaller inner tube diameter. Once satisfied that all is well and the seal made, seal the first side arm using "tack" or a cork. Hold the body to the rear of the flame and aim across the face of the hole while heating the flared "side arm" in the opposite side of the flame, when both faces are softening withdraw from the flame and quickly attach the "side arm", "run in" the seal while blowing alternatively down both the side arm and body. Once satisfied with the seal shape and the side arms are both in line with each other, remove the cork and warm the body/seal area until flecks of the sodium flame are seen, then allow to cool.

The condenser will now require annealing fully — to support the spiral in the annealing oven and stop it from sagging as it reaches annealing temperature half fill the body with common salt (Fig. 5.235). The spiral only has

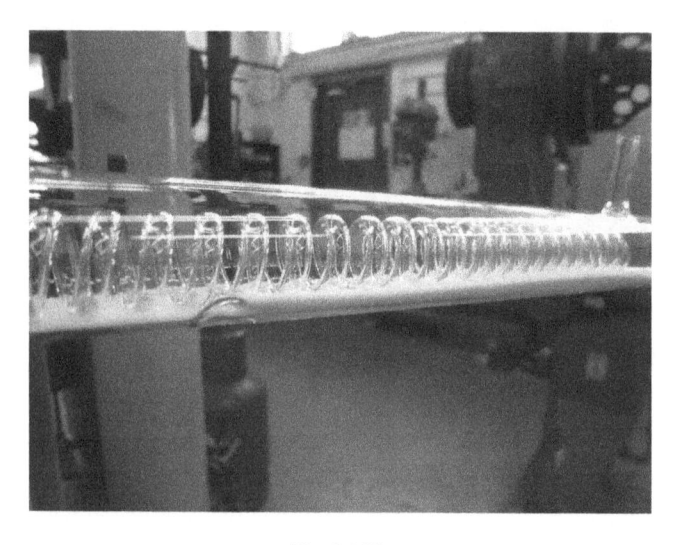

Fig. 5.235.

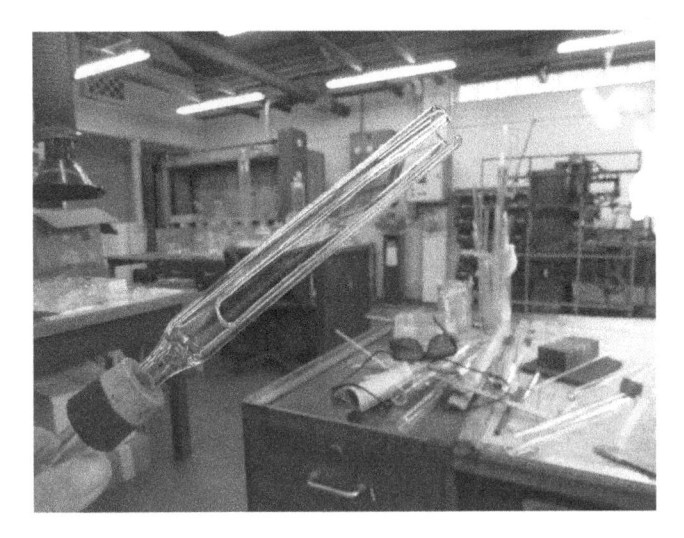

Fig. 5.236.

to be resting on the salt to give support. The salt can be washed away on removal from the annealing oven. Observe if you will the different types of spiral condenser, it will become obvious that the ends of the condensers have a variety of attachments, i.e. cones, sockets, flanges, ball joints and various other fittings. The narrower end of which would need to be fused to the body first, annealed and then the spiral or spirals sealed inside, various side arms made and the final end formed or attached. The longer or larger and heavier internal spirals will need far more sophisticated means of support whilst being made. The simplest would be a glass tube with a section cut away along its length which had been first prepared with an unequal join, the end of which is supported by a cork or rubber bung which enables quick and easy removal (Figs. 5.236–5.238).

This method of support can be modified in heat-resistant material or metal — all dependent on the number of condensers to be made. Larger heavier spirals are quite often given extra support by indentations formed in the outer jacket which are in close contact with the spiral at various points along its length so as to avoid undue sideways movement.

The larger coil condensers have support for the spiral built into the design so as not to have the side seals bear the whole weight of both the glass spiral and the weight of liquid passing through it.

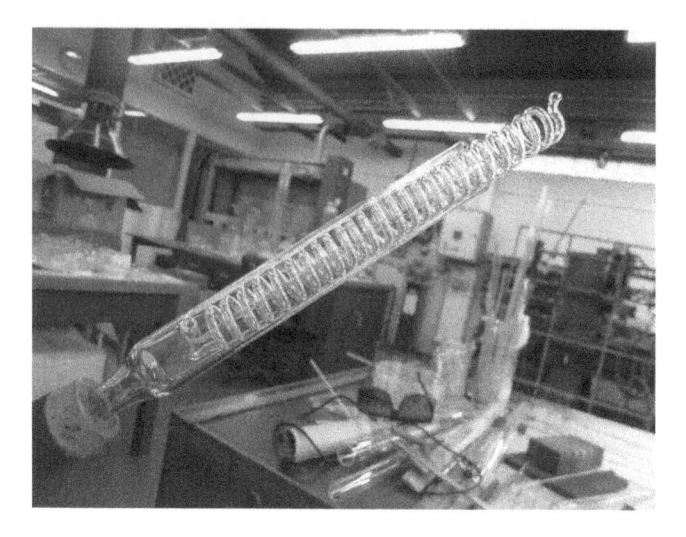

Fig. 5.237.

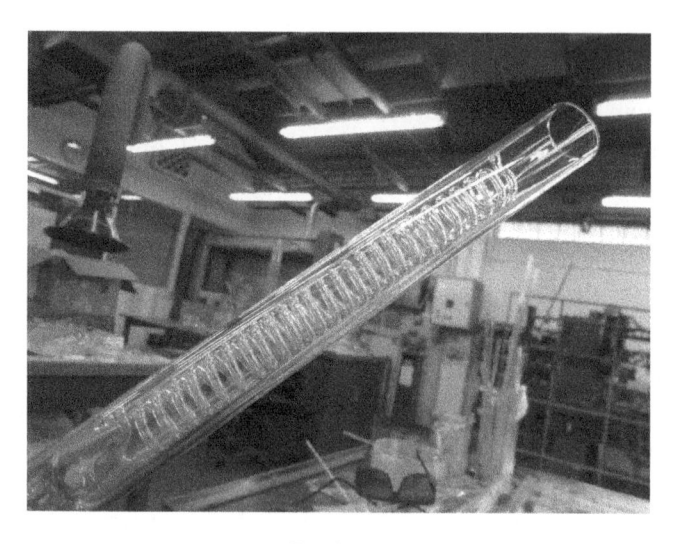

Fig. 5.238.

5.16 Capillary Joins

The definition of a capillary is a tube which has an internal diameter of hair-like thickness. Standard glass capillary tubing is supplied with a variety of both internal and external diameters. Specific internal or external sizes are

Fig. 5.239.

available on request although minimum quantities are requested by the manufacturer. Precision bore capillary tubing is available from specialist suppliers who resize capillaries by heat shrinking them onto a metal mandrel. This type of capillary is used where a known volume or flow is required, for example in the manufacture of UBBELOHDE and OSTWALD (Fig. 5.239) viscometers which are used to measure the viscosity of liquids. Both consist of a calibrated bulb, U bend and capillary insert.

Capillary joins introduce the use of both a blowing tube and a swivel (Fig. 5.240). The swivel will enable the capillary tube to be manipulated close to the flame which means that the glass can be blown while in the eyeline and also to avoid the blowing tube from kinking. The first exercise will use standard 8 mm ø tube before attempting the capillary join.

Prepare two pieces of 8 mm ø tube 25 cm long. Attach the blowing tube (rubber tubing 1.5 meter in length) to the swivel, then the swivel to the glass with a small length of heavy walled rubber tube (Fig. 5.241). The extra weight of the swivel and tube will be held in the left hand with the overhand grip. The second piece of tube will be held in the right hand from underneath after first sealing the end of the tube with either a small cork, rubber cap (Bobby) or a piece of tack. Bring the tubing edges together on either side of the 10 mm wide flame, the left hand rotating the tube constantly while a back and forth motion is sufficient with the right hand, as the edges melt

Fig. 5.240.

Fig. 5.241.

withdraw the glass from the flame towards you by 15 mm. Touch the edges together, then pull slightly apart and blow. It is relatively easy to control the blow as the glass is horizontal and directly in front of you. Do not overblow or allow the hands to wander apart, look for and attempt to control the wall

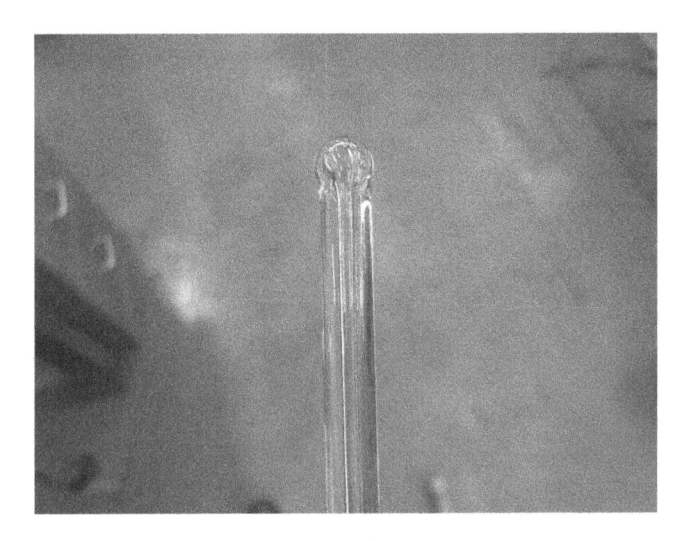

Fig. 5.242.

thickness. Rework the glass until satisfied that the initial join or ridge has been melted in sufficiently to make an even walled parallel join. Repeat the exercise until confidence has been gained to attempt the capillary join proper.

The capillary to standard 8 mm ø join: Place a 23 cm length of 1.5 mm bore × 8 mm ø capillary swivel and place in the left hand. Put the very tip of the tube into a 10 mm wide flame and watch as the end seals completely. Withdraw from the flame and blow to make a bulb (Fig. 5.242) within the wall of the capillary return to the flame and heat the very end of the tube while inclining the left hand. When it can be seen that the glass has started to soften withdraw from the flame by 15 mm and blow to produce a thin bulb (Fig. 5.243) which is then sliced off with the glass knife, if not successful at first repeat the operation until a sufficiently thin bulb is produced that can be removed. We are now ready to make join with a standard walled piece of 8 mm ø tube 20 cm long, one end of which has been sealed.

Place in the right hand and together with the capillary heat the ends of the glass in the edge of a 10 mm wide flame until molten. Withdraw from the flame and touch together, pull apart slightly and blow. Once the initial join has been made the control of the flame and its direction is of great importance due to the variation in wall thickness and the different speed at which the capillary and standard walled tubing melt. Reduce the size of the

Fig. 5.243.

Fig. 5.244.

flame to give a sharp flame 5 mm wide and concentrate the tip of the flame on what will become the shoulder of the capillary. Withdraw from the flame and continue to rotate the glass, blow to form a rounded shoulder (Fig. 5.244) at the join within the external diameter of both tubes.

To produce a tapered join pull slightly as the join is being blown. Repeat the exercise using two pieces of capillary 1.5 mm bore.

(1) Attempt to make the join simply by attaching the two cut ends as one would make a straight join.
(2) Blow a small bulb, then blow out each bulb prior to the join.
(3) Use a carbon reamer to open the capillary bore a little prior to the join. The carbon should be ground to a fine point using wet and carborundum paper. The flame is aimed at the glass and not the reamer if the carbon reamer is in contact with the hot flame for an extended period it will tend to "stick" to the molten glass and quite often break.
(4) Take two pieces of capillary with straight cut ends. Fit the swivel and seal one end, hold the edges together in line as you introduce the glass to the flame as the glass softens a slight pressure together will make a seal which can then be "run in".

Bends

In Sections 5.7 and 5.8 we have found that bends both right angle and "U" are readily made without blowing giving a bend with minimum disturbance to the capillary bore.

Repeat the exercises in Sections 5.7 and 5.8 using a variety of capillary bores and wall thicknesses.

Bulbs

Select a piece of capillary 1.5 mm bore 8 mm ø 23 cm long. Using both the swivel and the normal hand positions mark the centre of the tube and heat the glass in a soft glass, air oxygen flame as the glass softens the first blow will form an ellipse (Fig. 5.245) within the wall which when reheated increases in size to form a sphere without the need to thicken or gather the glass further. A bulb diameter three times the diameter of the capillary tube is a good guide to retain a reasonable wall thickness within the bulb (Fig. 5.246).

T pieces

This is not the easiest of exercises due to small internal dimensions and relatively thick walls. Take a piece of 1.5 mm bore capillary 20 cm long, attach

192   *Laboratory Scientific Glassblowing: A Practical Training Method*

Fig. 5.245.

Fig. 5.246.

the blowing tube and swivel, seal the open end of the capillary and mark the centre of the tube. Heat the glass at the mark in the hot gasses above a small hot oxygenated flame to prewarm the capillary so as to avoid thermal shock within the thick wall of the capillary. Hold the glass steady on the tip of the flame and blow as the glass starts to soften, watch as the capillary starts to

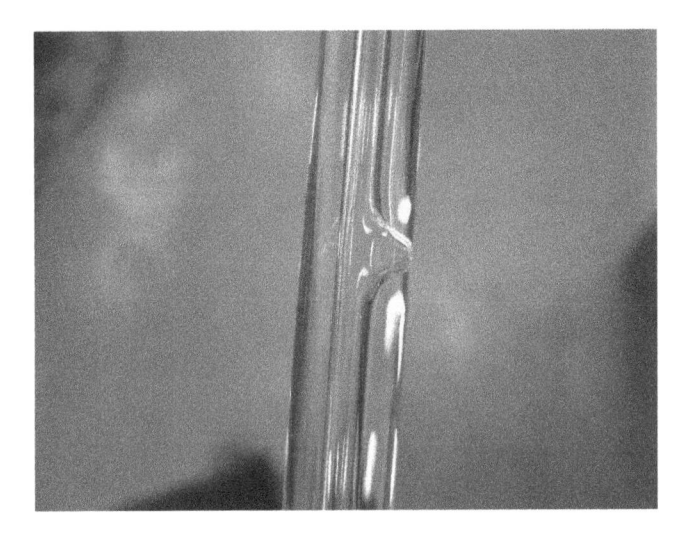

Fig. 5.247.

increase in size towards the flame, the offset bubble will eventually reach the surface (Fig. 5.247), at this point withdraw the glass from the flame and continue to blow to form a thin bubble which will be removed.

Alternatively do not withdraw from the flame, but continue to keep the glass in contact with the tip of the flame and continue to blow. The internal bubble will break through the glass wall and instantly form a hole.

Note: This method of making a hole will be used for standard walled tubing and adapted further as the work becomes more complex.

Depending on the size of the capillary arm, we are about to attach that will govern how we prepare the side arm, either join a straight cut edge, a blown out bubble or use a carbon reamer to flair out the end of the capillary. The smaller the bore the greater the need to increase the size of the bore to be joined. Seal on end of the side arm. To help make the join in such small bore tubing it is advisable to use Didymium glasses to help see through the sodium flame. Hold the longer tube to the rear of the flame and the side arm to the front and right of the flame, when both edges are hot withdraw from the flame, touch together and blow. Run in the join using as small a flame as possible on either side of the join while using the cooler opposite side of the join to give support. The exercise will need to be practised repeatedly.

Fig. 5.248.

The final outcome will be that at the very least a "T" piece or side arm in standard tubing will improve beyond measure.

On many occasions there has been a need for a flat ended capillary. One method is to first blow a bulb within the wall of the capillary (Fig. 5.248), then while holding the capillary at an angle of 45° play the flame at the very end of the capillary tubing while rotating constantly, watch carefully as the bulb melts back to give a flat internal surface and withdraw from the flame before the capillary itself constricts. (Fig. 5.249)

Bore/Size/Measurement

(1) *Optical graticule*: The end of the capillary is placed against a precise scale which is viewed through an eyepiece where the bore can be read directly from the scale.
(2) *Internal sizing gauge*: A precise metal point is placed inside the capillary. The gauge then makes contact with the end of the capillary tube and bore read directly from a vernia scale.
   The above tools are robust enough to be used at the bench.
(3) Optical and digital microscopes are available for absolute precision when required.

Fig. 5.249.

Fig. 5.250.

The spike that is often formed after snap cutting a piece of capillary tube must be removed before handling or blowing. Under no circumstances consider putting something so dangerous in your mouth to blow. (Fig. 5.250)

5.17 Repeat — Alternative Techniques

T Pieces

The preparation for the side arm hole so far has been fully blown out, the thin bubble then sliced off. An alternative being as follows: take a piece of 10 mm ø tube 25 cm long, mark the centre and seal one end. For the first exercise we will attach a blowing tube to the open end, this will ensure that the hole we are about to make will be a little farther away from our face!

Aim a small flame at the centre mark, once the glass has started to soften withdraw from the flame and give a controlled blow to produce a dome, hold the tube to the left of the flame and to the rear. Once the glass has started to soften withdraw from the flame and give a controlled blow to produce a dome. Return the glass to the flame which should touch the top of the dome. Blow while still in contact with the flame. The glass will soften and start to form the thin bubble but will burst open and immediately melt back to form a hole (Fig. 5.251). We have now produced a hole complete with a slight shoulder without any loss of wall thickness. When the side arm join is made both the side arm and proud hole will melt evenly so that the initial join will be far easier to complete. Once the initial join has been made reduce the size of the flame and aim directly into the corner

Fig. 5.251.

of the join where the wall thickness of both pieces will be similar and should flow together evenly.

To gain experience use the blowing tube to create a hole in a variety of tubing diameters. In time a blow directly into the tube can be practised. This technique can be used in a variety of situations. For example — when making a join to a round bottomed flask — as the flask wall is relatively thin and can ill afford to have its wall blown out and reduced further. One can "blow out" the hole at the flame face and then increase the size or shape of the hole using a carbon/graphite faceted reamer. This will help retain the wall thickness at the join.

A valuable exercise in "controlled blowing" will be to take a piece of 20 mm ø tubing 30 cm in length and mark two parallel lines along its length using the edge of the bench as a guide (Fig. 5.252). Make a mark 10 cm from one end and seal the tube with a cork or similar. Using a small flame heat the tube at the 10 cm mark, withdraw from the flame and blow to make a small dome, move 15 mm along and blow a 2nd dome then a 3rd, 4th and 5th (Fig. 5.253). The marked line burns off when heated but the untouched cooler parts will still act as a guide. At this point if a polarising strain viewer is available it will be interesting to see just how much "strain" is present in the

Fig. 5.252.

198   *Laboratory Scientific Glassblowing: A Practical Training Method*

Fig. 5.253.

Fig. 5.254.

glass (Fig. 5.254), if on the other hand a strain viewer is not available be aware that there will be a great deal of heat-induced "strain" present — sufficient for the glass to "stress relieve" itself by first cracking and possibly falling apart. Therefore we will have to balance the temperature of the glass and defuse the

Fig. 5.255.

strain by warming the whole of the five inline domes in a feather-like flame until specs of the orange sodium flame start to appear, which will indicate that the glass is now above the strain point and the heat/strain is defused, i.e. flame annealed. Allow the glass to cool without being in contact with a cold surface.

The task is now to make a second set of five domes in line with the first on the opposite side of the tube. On completion "flame anneal" both the first and second lines of domes. Allow the glass to cool completely and then inspect the work with a critical eye (Fig. 5.255).

Are both lines parallel?
Are the domes evenly spaced?
Are the domes of similar shape?
Are the domes of similar height?

Repeat this exercise and concentrate on making a piece with inline domes of similar shape and spacing.

Within this book there is a colour photograph of a "swinging bucket still head" with four parallel lines of small domes all at a set distance apart.

Look closely at the photograph, learn, practise and aspire (Figs. 5.256(a) and 5.256(b)).

(a)   (b)

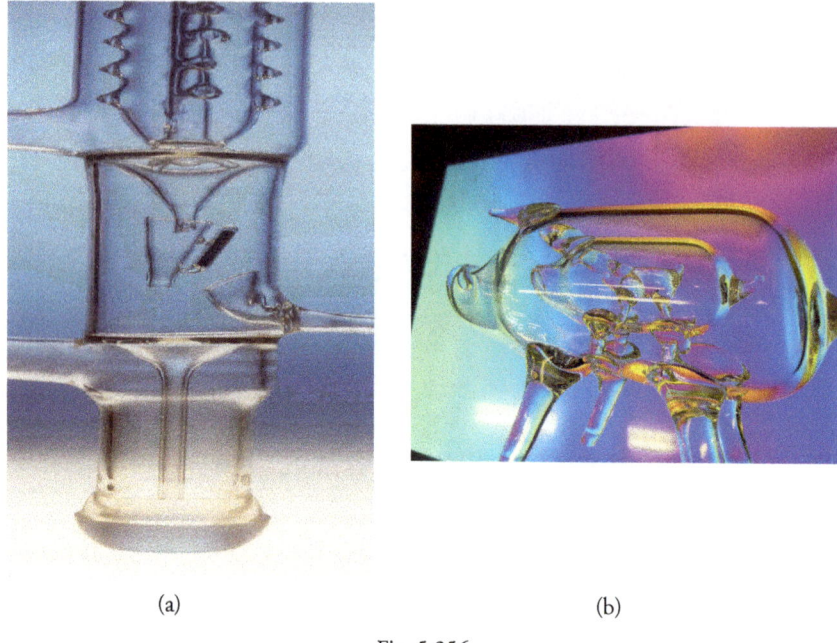

Fig. 5.256.

Supported Internal Seals

We have to date used corrugated paper to give support to the internal seal. The advantage being that it is cheap, easy to cut to shape, gives good support and is quite easy to remove. The disadvantage being that it is combustible so the support given is restricted by the size of tubing as larger diameter tubing requires a corresponding increase in the amount of heat needed to melt the glass which tends to "roast" the corrugated paper the vapours from which can contaminate any seal being made.

I will mention in passing "Asbestos Paper!" An ideal material in so many ways as it gave support at high temperatures, but the fibres are DEADLY. There are severe restrictions as to its use worldwide and SHOULD NOT BE USED under any circumstance. Woven fibre glass can be used as a support material although it is a little smooth if not slippery sometimes causing the supported seal to move.

Like fibre glass alumina silicate paper has its uses as a support material as it will withstand high temperatures and can be easily cut and shaped.

The disadvantage is that once heated the bonding agents holding the fibres together outgas leaving only the fibres which when removed are unsuitable for reuse. There is therefore a cost element to be considered to its use.

Corrugated Graphite Tape

Although there is a cost aspect the material is clean, easy to use, and gives good support. Removal of any type of packing has to be thought through carefully as is the sequence of manufacture of an item with multiple internals, leaving graphite tape, alumina silicate paper or glass fibre inside even a partially made item is not an option, a means of extracting the support material must be considered a priority in first designing an item of multi-walled glassware. The above materials are virtually impossible to be removed chemically.

Although copper wire made into spirals will give support to seals. The wire is first wrapped around a rod of similar diameter to the annular space between the outside diameter of the tube to be supported and the internal diameter of the outer tube. After winding the copper spiral the wire is "stretched" a little so that there remains some "spring" within the copper wire as it gives support to the inner tube. The wire can be left inside the annular space after the seals are made and removed later with concentrated nitric acid. One would have to complete this operation in a fume cupboard while conforming to all local safety, Personal Protection Equipment and disposal regulations. The copper wire will dissolve in the acid. I am aware that "aspirin" tablets have been used to give support to an annular space. They are easily shaped and even joined together if moistened to give the required thickness and are easily removed with water.

There has been a continuous search for an alternative to asbestos as a support material. One of the more recent materials that I have become aware of that is resistant to heat which grips the glass well and is easily cut to shape is heat-resistant woven cloth, which is used for repairing car exhausts and can be folded on itself like paper and even cut with scissors, most importantly it is reusable.

Various support jigs are available in the market, their use makes supporting internals ideal in a production environment where multiple units are

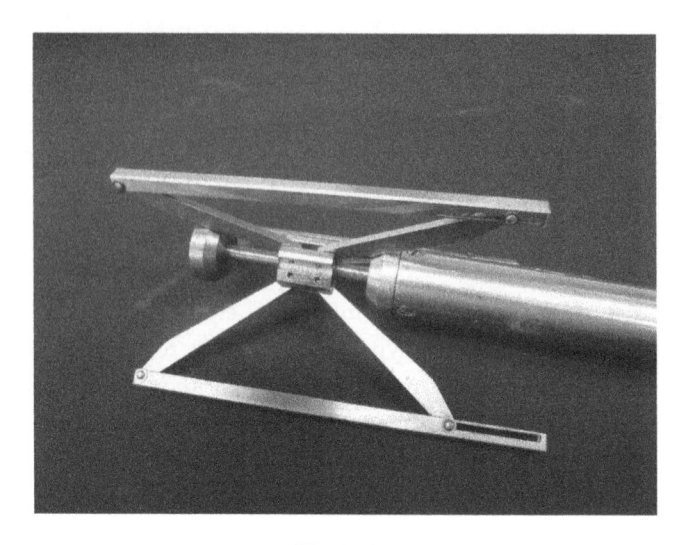

Fig. 5.257.

being produced, although initially very expensive one would have to consider the cost-effective aspect of each application (Fig. 5.257). Similarly double back to back lathe chucks will give support to specific types of work. When using metal jigs to support internal components it is essential that the jig is warmed prior to starting as the difference in expansion rates between the glass and metal will cause catastrophic collapse if the metal is not pre-warmed.

Let us retrace out steps a little and use the corrugated paper support method once more while using the suggested sizes of tubing outlined in Section 5.10:

D1 38 mm ø   L1 220 mm
D2 16 mm ø   L2 180 mm

Now 2 × side arms 16 mm ø L1 20 mm flange one side arm to approximately 22 mm ø.

Support the inner tube using corrugated paper and proceed to the point where we have successfully made the seal (Fig. 5.258). Heat the 38 mm ø tube at the shoulder and blow to create an overblown domed end, allow this to cool slightly, then reheat and blow out the centre of the seal, place a cork in the open end of the 38 mm ø tube, concentrate the flame across the open

Fig. 5.258.

Fig. 5.259.

end of the seal until the blown out ridge melts back so that there is a smooth join between the inner and outer tubing, we have in effect made a "Dewar" seal (Fig. 5.259). Take the pre-prepared 16 mm ø tubing in the right hand and heat both the Dewar seal and flanged end in a hot flame which is large enough to heat both edges of the glass. Heat the glass evenly with the left hand constantly rotating in one direction while the right hand rotates back and forth until the glass is quite hot but before the seal collapses. Withdraw from the flame a short distance and touch the edges together, pull apart slightly and blow. If successful, one will see that the seal is both thick and robust.

With this type of seal "timing is everything".

Rotate and straighten the seal whilst still hot, create a hole close to the shoulder and attach the side arm.

Second Exercise

Using the same size and length of tubing as used for the previous exercise support the 16 mm ø inside the open ended 38 mm ø tube allowing the 16 mm ø tube to protrude beyond the edge of the outer. The opposite end of the 16 mm ø tube must be sealed with a cork.

The aim now is to ream (or flange out) the internal tube until it touches the outer (Fig. 5.260). To do this we can use either a carbon rod the end of which has been ground to produce a long faceted taper, alternatively a wooden handled brass reamer which will have to be lubricated with either beeswax or a proprietary graphite/beeswax mix.

Fig. 5.260.

Fig. 5.261.

If using the brass reamer we need to first warm it in the flame then the reamer is touched upon the beeswax which will melt giving a protective coating to the metal. Hold the glass in the left hand and rotate the glass evenly in the flame concentrate on the end of the inner tube as the glass softens, place the tip of the reamer inside and rest the point on the solid glass well in from the opening. Incline the left hand slightly together with the right and watch as the inner tube folds up while in contact with the reamer. Try as far as possible to keep the reamer out of the flame while now concentrating the flame on the shoulder as the inner tube is brought in contact with the outer.

A little blow will confirm that the seal has been made. Return to the horizontal position and have the flame directed across the face of the seal, withdraw from the flame and blow to finalise (Fig. 5.261). One may see that the internal tube's diameter has constricted when the blow was made. This can be rectified by heating once more and using the reamer in a horizontal position to open out the inner tube (Fig. 5.262). A carbon rod would be easier to use in this instance by placing it well inside the tube and resting it on the cooler glass to act as a guide for the required internal diameter (Fig. 5.263). Alternatively, better flame control will avoid this happening, always be aware of the "edge" of the flame.

Fig. 5.262.

Fig. 5.263.

Coil Winding

Larger coils of either overall diameter or length are best made on a slow running lathe or single head chuck. Most lathes have high torque and very poor low speed control as they tend to surge unless geared specifically for low

Fig. 5.264.

speed spiral winding. Fortunately retro-fit low speed drives are available (Fig. 5.264). The need for smooth low speed control is paramount during the coiling process, ideally a remote speed control and emergency stop should be as close as possible to the operator, should the glass be heated unevenly and become rigid while still rotating!

A burner giving a large feather-like flame is required. The work area must be free from obstructions especially when longer lengths of tubing are to be manipulated. If a slow running attachment is available the head stock must be detached from the main drive shaft via a clutch. The head stock and chuck will then be free running, driven only by the low gear external drive.

Standard lengths of tubing will need to be joined together prior to the work starting, alternatively standard lengths can be joined as work progresses although this will have to involve a second person to make the join and may well require the spiralling process to be paused, whichever method used the join will be made with a hand torch and blowing tube. If one has an assistant and no slow running attachment is available one person can rotate the free running chuck head by hand while the second makes the spiral … It is possible but will require a great deal of understanding and cooperation between each person.

For the single person complete with a low running lathe the procedure will be to ensure that the speed control is set at zero. The mandrel set in place and the chuck tightened. The burner flame set to give a large feather-like flame. Prewarm the mandrel, preshape the end of the glass so that it will hook into the mandrel. There are two methods to start. First — hook the glass to the mandrel and play the flame along the top of the glass, then as the glass is seen to soften by sagging start the motor at the lower speed and watch as the glass is taken over the top of the mandrel, the flame should be kept in contact along the top of the glass. Second — have the lathe in motion, pre-warm the glass — balancing the temperature with the rotation so that when the hole in the mandrel appears once more the warmed glass can be hooked into the hole and the flame immediately played along the already softening tube.

Once more timing is everything as the pre-warmed tubing must be about to sag as it is hooked onto the mandrel. Once the first turn of the coil is completed the temperature and position of the flame can be adjusted and possibly the speed of rotation increased a little if felt necessary to balance the softening glass. The glass passing through the flame and over the mandrel is opposite to the bench method where the glass is drawn around the mandrel from underneath (Fig. 5.265).

Longer spirals require longer lengths of tubing if the size of the workshop allows join the lengths of tubing beforehand and support the tubing on rollers as it is fed onto the mandrel. Alternatively make spirals, then join the spiral sections together while holding the sections in the jaws of the

Fig. 5.265.

Fig. 5.266.

Fig. 5.267.

lathe, use a hand torch to make the join; this will help keep everything inline (Fig. 5.266). If one of the chucks is disconnected from the drive it will enable you to align the spiral in and out and back and forth while completing the join. In many ways this will be dependent upon the pitch and gap between each coil (Fig. 5.267).

Bulb Blowing

Large scale manufacture of round bottomed flasks are mould blown from the molten state. Whether automated or by hand the first part of the process after the gathering of the molten glass is the "paraison" or pre-form prior to blowing. Larger flasks are blown into hinged two part moulds the inside of which is often moistened prior to blowing. The moisture immediately turns to steam so the glass surface is shaped against a layer of steam rather than the mould itself, so reducing the possibility of mould marks.

At the laboratory scale it is quite normal to make round bottomed flasks from pre-drawn tubing, although there are occasionally circumstances where specific size flasks are required, for example "sugar flasks" ranging in size from 25 mL to 200 mL capacity, which are made from soda-lime glass, often pre-shaping the flasks are blown into spring loaded split moulds. Soda glass is far more amenable to this type of work whereas borosilicate glass is less so as it has a far shorter working time. Borosilicate can be blown into "half moulds" consisting of a block of carbon/graphite with a semi-spherical shape of the required size cut into the block, into which a hot pre-shaped tube is rotated and blown. Another method that can be tried is to use a semi-circular carbon former against which tubing can be blown whilst being held in a lathe to "finish off" where a specific diameter of flask is required repeatedly.

Free blown flasks either on the lathe or at the bench are more in tune with a research glassblowing laboratory rather than a production line situation. Bear in mind that round bottomed flasks are designed to "contain" a working head space above the liquid level rather than deliver a specific volume, a close inspection of manufacturer's diameter volume specifications will confirm this by comparison with absolute sizes.

Indicated Volume (mL)	Working Diameter (mm)	Absolute Diameter (mm)
50	51	46
100	64	58
250	87	78
500	105	98
1000	131	124
2000	167	156
3000	189	179

(Continued)

(*Continued*)

Indicated Volume (mL)	Working Diameter (mm)	Absolute Diameter (mm)
4000	205	197
5000	222	212
6000	235	225
10000	290	267
20000	365	337

5.18 Cones, Sockets, Spherical Joints, Screw Joints, Flat Flanges and Buttress Joints — Their Application — The Use of Mechanical Aids Such as Lathes, Cut-Off Machines, Linisher Belts Diamond and Carborundum Lapping Machines

For centuries, closures for glass vessels such as cork or wood have been used to prevent the ingress of air into wine bottles. The neck needs only to be consistent in size with a robust rim. I enclose a series of engravings taken from the frontispiece of three books by Joseph Priestley on his work to define various Airs in 1777 (Drawings 5.1–5.3).

Through experimentation he had clearly defined the main constituents of the air we breathe, part of which he defined as dephlogisticated air, it was the French Chemist Antoine Lavoisier who termed the name oxygen.

Joseph Priestley like Lavoisier repeated experiments to prove or disprove a chemical theory. They used glass to conduct their experiments so as to be able to witness any colour change or chemical reaction within the clear glass vessels. This approach continues to the present day. View carefully the spherical vessel which has a side arm to keep a set level within the flask, a bottom run off and the jointed fittings to introduce new material.

The glassware was far removed from the retort (Fig. 5.268) and alembic used by alchemists. The glassware shown I believe was the beginning of "Scientific Glassblowing". The early type of socket is clearly seen on a "kipps" apparatus (Fig. 5.269), the joint has been formed from the molten state the hole has been ground out to a 1:10 taper (Fig. 5.270). This 1:10 taper is the basis for all interchangeable cones and sockets of today. Scientific glassware requires that cones and sockets be made to international standards to guarantee interchangeability with good gas and liquid seal through which liquids and gasses may pass.

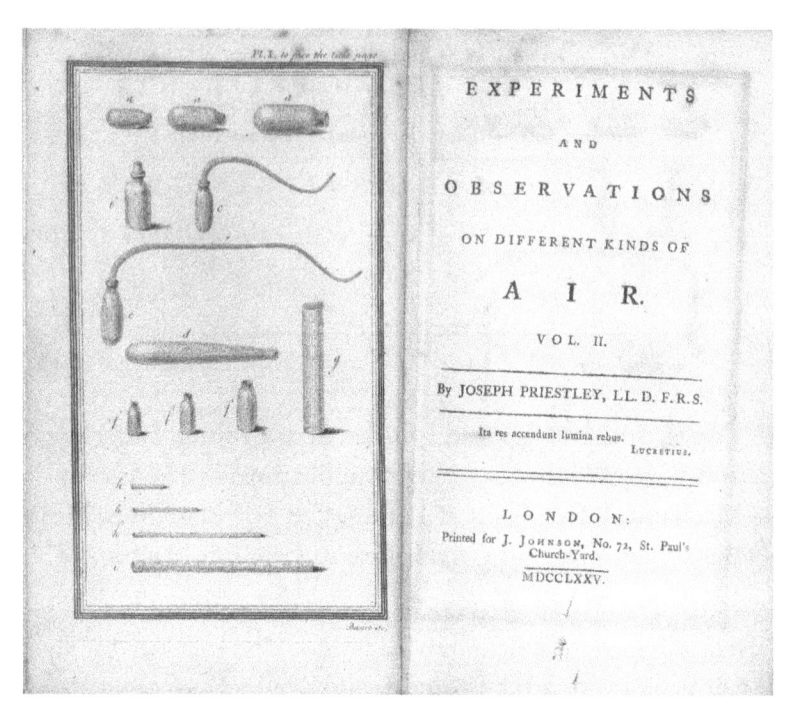

Drawing 5.1.

Although cones and sockets conform to the 1:10 taper there are four variations in the lengths of the taper for both cones and sockets.

Size A — long preferred size in USA.

A19/38	large diameter	18.8 mm
	small diameter	15.0 mm
	Length of ground surface	38.0 mm
B19/26	large diameter	18.8 mm
	small diameter	16.2 mm
	Length of ground surface	26.0 mm
C19/17	large diameter	18.8 mm
	small diameter	17.1 mm
	Length of ground surface	17.0 mm
D19/9	large diameter	18.8 mm
	small diameter	17.9 mm
	Length of ground surface	9.0 mm

Drawing 5.2.

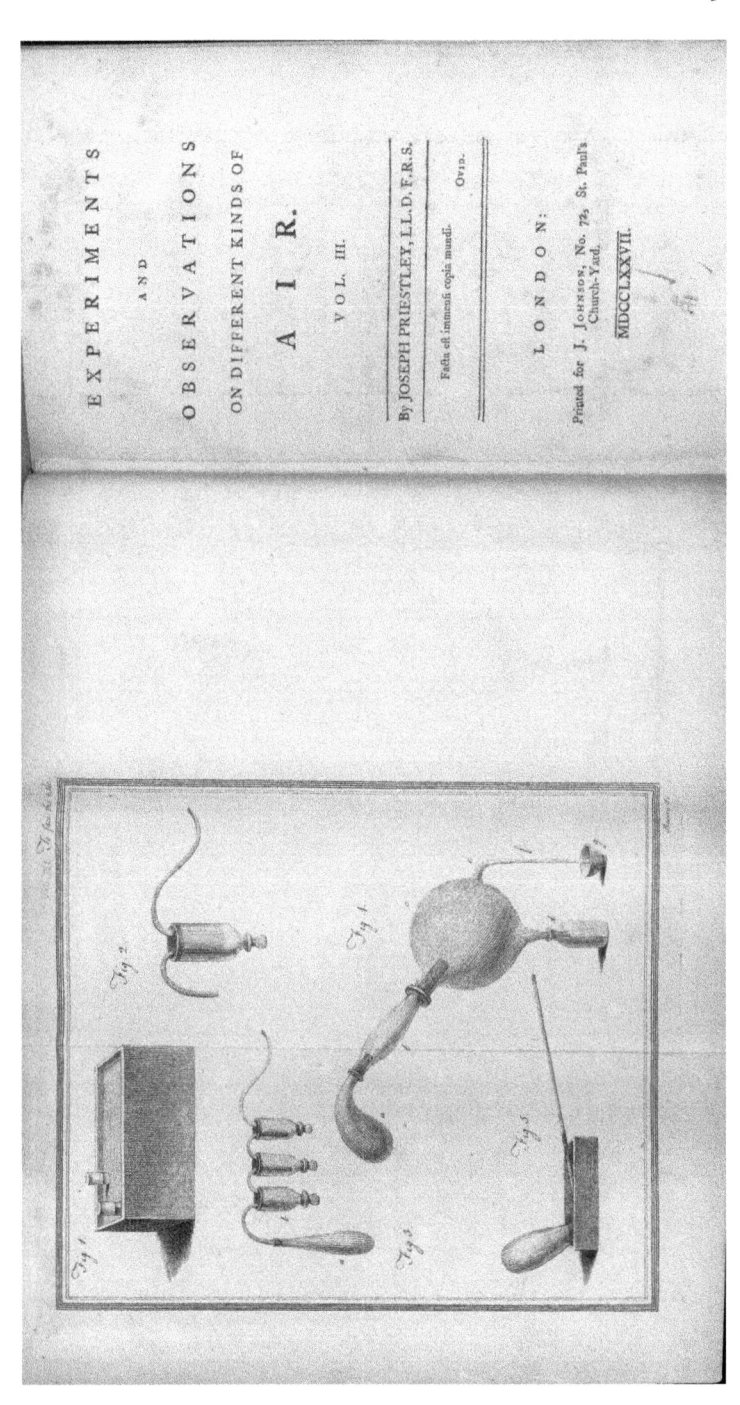

Drawing 5.3.

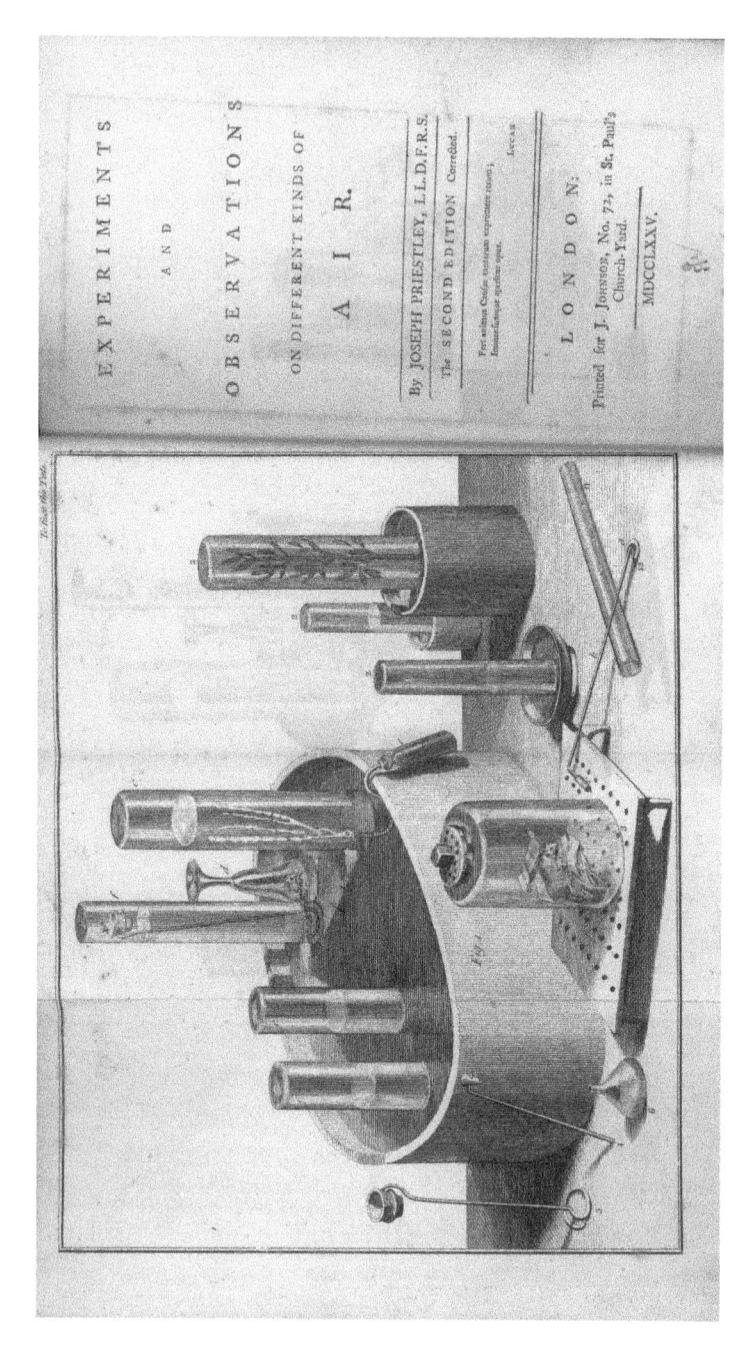

Fig. 5.268.

Fig. 5.269.

Fig. 5.270.

The B length cones and sockets are the preferred size in Europe. It will be seen that the first number defines the nominal large diameter and the second the length. In use cones and sockets are referred to with just the letter and first number. Skirted cones have an extension beyond the narrow end of the cone which can be cut or ground diagonally to produce a "drip cone", as the name suggests the diagonal shape will encourage condensing liquids to drip from one specific point which in itself will encourage further liquid to gather at that point (Fig. 5.271). The skirted cone can also be extended further by making a straight join directly to the skirt without the threat of distorting the cone (Fig. 5.272). The length of the extension required is dependent on the needs of the end user and the type of chemistry being undertaken, for example the extension can be either below a liquid level or above.

Cup or Extended Sockets

Cup or extended sockets are often used when there is a possibility of spillage in use. From a purely glassblowing point of view extended sockets are used when making a jacketed column Dewar-type seal, where a socket is required to extend a column to accommodate a still head or condenser, for example.

Fig. 5.271.

Fig. 5.272.

Once more then Dewar seal can be made with less chance of distorting the socket although good flame control is required (Fig. 5.273).

A further development of the standard cone and socket has been the addition of a threaded locking system to the socket which when combined with an "O" ring retention ridge formed on the shank in combination

Fig. 5.273.

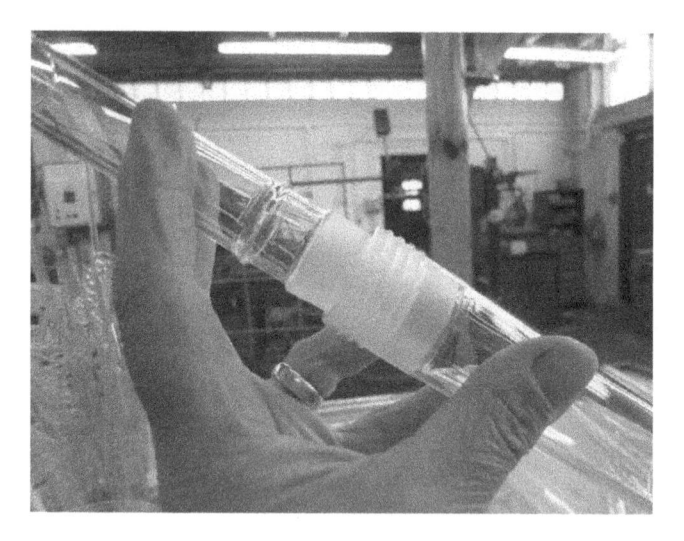

Fig. 5.274.

with various threaded retention caps and seals enables the joints to be used under both vacuum and pressure with confidence. The ground surfaces of both the cone and socket conform to the standard "B" lengths and diameters (Fig. 5.274).

Refer to the local manufacturers' catalogue for the full range sizes of cones and sockets currently available.

Spherical joints = Flexibility

Although glass is an adaptable material it has its limitations especially when it comes to physical stress — strong in compression — weak in tension. Misaligned joints can put a great deal of stress on glass constructions, it is important to include some flexibility into a system where there are multiple clamps holding manifolds and trap systems in line and in place.

Spherical joints, either greased or grease free will give such flexibility by taking up potential misalignment whether intentional or not. The greased spherical joints consist of a ball and socket ground to a standard radius which are held together with a custom-made spring clip (Fig. 5.275).

The numbering system for the size code is the outside diameter in millimetres of the ball and the inner diameter at the lip of the socket where they are at the maximum. The grease free ball and socket consists of an unground ball which has a ridge to accommodate a PTFE-coated neoprene "O" ring seal. The socket has a polished inner surface against which the O ring makes contact, a sprung metal clip holds the ball joints together tightly once in alignment.

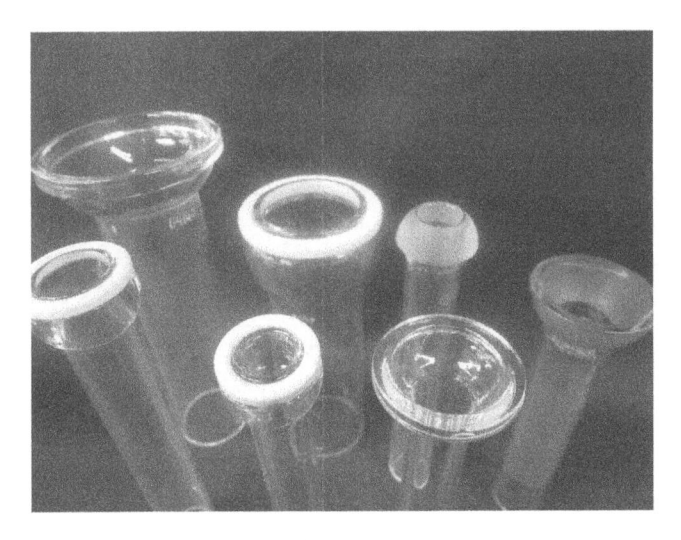

Fig. 5.275.

There is a wide selection from various manufacturers of screw fittings, sadly not all are interchangeable at the time of writing. The size code is defined by the external diameter of the shank. PTFE and silicone rubber washers are used to make the inner seal while a threaded plastic cap makes the connection with the glass. Consult the local manufacturers' catalogue for specific size and range.

Flat flanges and buttress joints provide a larger diameter entrance to a vessel and in turn a larger lid closure which enables a number of inlet and outlet ports for stirrers and thermocouple pockets to access a reaction vessel (Figs. 5.276 and 5.277). The bore of the shank from which the flat flange is made defines the size. Buttress joints are press moulded either as a complete lid to which a variety of joints can be attached. Buttress joints are also available with a short section of tubing attached to join to a vessel. The faces of the buttresses are ground flat and can be used face to face, the seal being made with grease. They are designed to be used grease free as buttresses are also available with a groove set in the front face into which an "O" ring fits. The "O" rings are made of either silicone rubber or have a silicone core with a seamless FEP coating (tetrafluoroethylene hexafluoropropylene copolymer).

Note: The maximum temperature in use is 230°C. The chemical resistance is equal to PTFE. Two buttresses are held together with a profile spring clip.

Fig. 5.276.

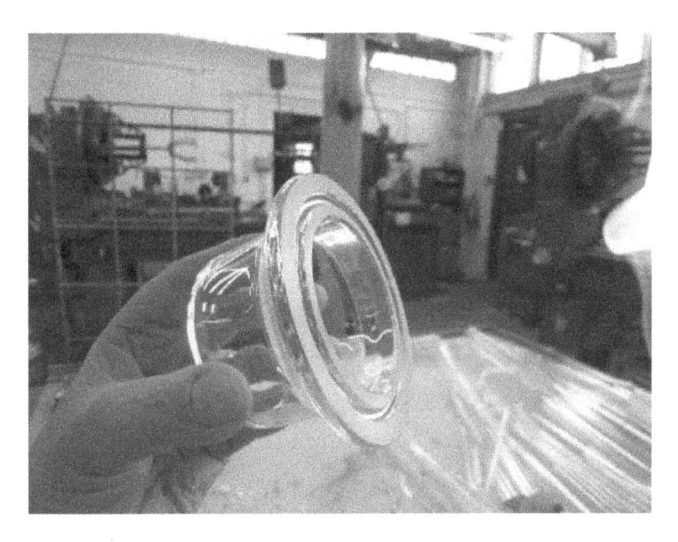

Fig. 5.277.

A further point to note:

When using PTFE (polytetra fluroethane) as a sealing agent either in the form of an "O" ring seal or flat washer the maximum in use temperature is 300°C, above which the material produces (perfluroiso butyler) fumes which are extremely toxic!

For this reason, also remember never to leave a PTFE key in a tap when glassware is placed in an annealing oven.

Cut-Off Machines

Use high speed cutting discs the periphery of which have various grades of industrial diamond impregnated in a phosphor bronze rim the thickness of which is greater than the thickness of the disc which gives clearance to the cutting action. There are two types of cut-off machine in common use where the cutting disc is positioned either above the work or from underneath. The cutting action of a diamond impregnated disc is to grind through the glass causing what amounts to glass dust particles which are immediately flushed away to drain via a sedimentary trap whereas in a recirculatory system there is an accumulation of glass particles which are abrasive to pumps and seals. When working close to the point at which the

Fig. 5.278.

cutting is taking place one may be splashed or sprayed with cooling liquid, so better for it to be water rather than the soluble cutting fluids used in recirculating systems. The glass rests upon a sliding bed which will have either a screw drive or push slide to traverse the glass through the blade or an option to use either method so that the glass when placed against an end stop can be "felt" as the high speed disc cuts or grinds its way through the glass (Fig. 5.278).

A guide for the width of the cut can be attached to the end stop so that the width of cut can be measured from this point. The guide will give support to the glass as the cut progresses. When cutting across the diameter of the tubing cut through just less than half way, withdraw the glass from the disc and rotate the glass so that the leading edge of the blade is between 90° and 45° to the cut edge of the tubing until completely through, especially when cutting larger diameters. Whatever size of tubing to be cut the final few millimetres of the cut is critical. It is essential to use a light touch to allow the wheel to gently grind through rather than force the wheel which will inevitably cause chipping at the final edge of the cut.

When tubing is to be cut along its length or a slice of tubing is to be removed a simple cradle can be used. The height of the blade is raised enabling a cut to be made through one thickness of glass.

Fig. 5.279.

One could use the side gauge, but having a cradle where the tube sits within a 90° bed is far more stable (Fig. 5.279). Free hand cutting is possible although I would not recommend it, preferring at all times to have the glass resting upon the back support when the glass is in contact with the blade, especially when cutting flat glass.

A simple support can be made when making an angled cut. Refer to Chapter 12 where Gary Coyne describes his jig and the material used in supporting an angled cut. There are available custom-made angle cutting attachments specific to the glass cutting equipment in use.

Flat lapping machines for the grinding of flanges use either various grades of carborundum powders mixed with water on a rotating cast iron disc. A more recent development is the use of industrial diamond held within polymer beads on a backing material which are rotated at high speed whilst being cooled with water. The flat surface of the lapping machine enables large flange surfaces to be ground so that two flanges can be clamped together to provide a vacuum or pressure tight seal on a vessel. The faces of the two flanges are lubricated with a suitable grease.

The diamond lapping machines run at high speed, great care should be taken when first making contact with the glass. Have the machine in motion, then lightly touch the glass onto the surface of the diamond disc. DO NOT

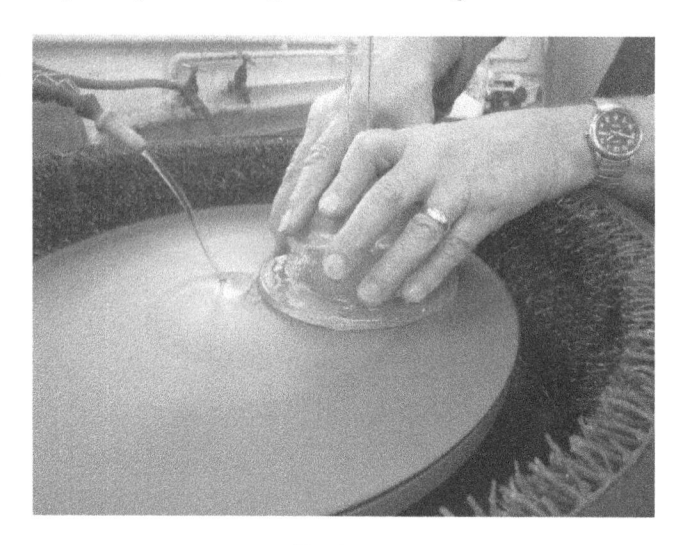

Fig. 5.280.

have the glass in contact with the disc, then attempt to start the machine as the take-up is sudden and there is a good likelihood of the disc "catching" the glass (Fig. 5.280).

Various grades of disc are held in position with hook and loop closures for quick replacement with a finer grade (Fig. 5.281). The carborundum lapping machine runs at a slower speed and uses various grades of carborundum powders which are made into a slurry with water directly onto the rotating cast iron disc.

The different grades of carborundum have to be washed off both the glass and the disc between each grade. The carborundum ground surface gives a good finish for the sealing greases to adhere to. If a lapping machine is not available a piece of "float" glass can be used as a grinding surface, bearing in mind that it will have to be replaced often as both flange and plate are in contact with the abrasive. If using a glass plate to grind a flange place a small amount of carborundum on the plate, add sufficient water to make a slurry and use the flange itself to spread the carborundum over the plate using a figure of eight motion. Watch the flange surface constantly until an even ground surface is seen, then wash everything and repeat using a progressively finer grade of abrasive.

Keep each grade of carborundum separate in sealed containers. Fine grades of carborundum contaminated with coarser grades render them unusable.

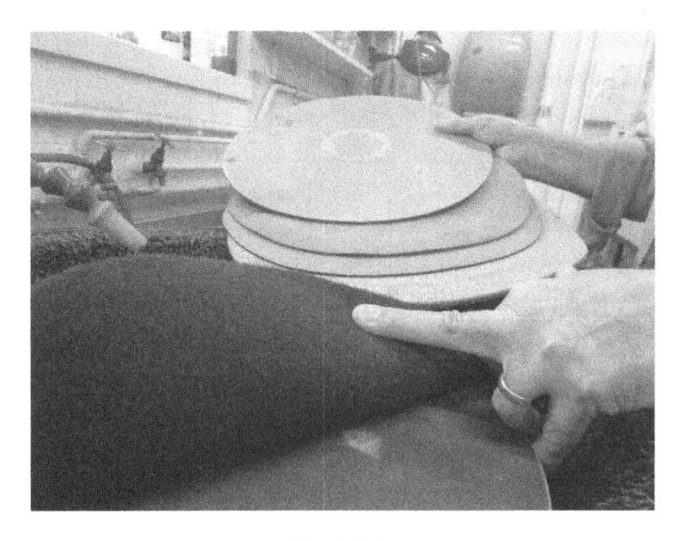

Fig. 5.281.

Linisher Belt Grinding

A water cooled vertical continuous belt running at high speed coated with various grades of industrial diamond held in polymer beads will reshape and rapidly remove glass when using the coarser grades. The surface finish will improve with the finer grade linishing belts until eventually polishing grades will produce a fine clear surface. It is essential that each graded step is completed and that no intermediate grade is missed or left out as this will lead to scratch marks on the surface of the glass.

A support for the glass is set at right angles to the belt. Sintered and optical discs can be resized using a suitable attachment which will be covered in detail in Section 5.19.

Whenever one is dealing with glass in any shape or form that is in contact with rotating, cutting or grinding machinery it is most important to have a safety cut-off switch for the power, either a pressure switch on the side of the machine or a floor mounted foot switch that can be operated in an emergency when the hands are otherwise occupied.

Lathes of various types and sizes have their own dangers, not only rotating machinery alone but also by having glass clamped between two rotating chucks whilst being heated and reshaped, it is essential to work methodically without cutting corners or daydreaming.

Be aware of the material and concentrate at all times. When replacing or removing glass from the jaws of a lathe it is best to first STOP the lathe rotating and remove the glass from the lathe after working with tongs.

A rotating piece of newly cut glass is no place for unprotected fingers. The consequence of such action can be life changing. Following on from the lathe made straight join in Section 5.11 I will describe the basics of lathe-made unequal joins. Lathe techniques follow those for bench work, the main difference being that on the bench the hot glass can be inclined to help form a shoulder or sphere as the glass when hot can be encouraged to flow into the join.

Lathe work is horizontal which means that shoulders have to be prepared in a way that avoids the loss of wall thickness by either heating the tube then using a rounded carbon rod to make an indent in the larger diameter to match a smaller. If the diameters are vastly different the reduction in diameter will have to be gradual and made in stages as the burner is moved along the tube to remelt and retain wall thickness by a slight pushing together before further tooling to reduce the diameter can proceed once more. Once the required diameter is gained the options are as follows:

(1) The glass can be allowed to cool and cut on a diamond wheel as the required reduced diameter.
(2) The constriction heated further and tooled once more with the carbon rod until the tubing is sealed, the end of which is then reheated and blown out to create a thin bubble which can be sliced off in much the same way as on the bench.
(3) The constriction is heated with a narrow fierce flame as the glass softens the right-hand chuck is moved further to the right while the flame remains static and in contact with the glass while a constant blow into the glass is made via the blowing tube and swivel. The glass will thin at the flame edge until it melts through, continue to move the right-hand chuck away as the flame makes a neat edge to the newly created hole without the remnants of any bubble floating and sticking to the lathe bed.

If the two sizes of tubing to be joined are not too dissimilar the end of the larger tubing can be heated and tooled down to the required diameter (Fig. 5.282) either free hand using a carbon paddle or a carbon plate secured on the tooling bar. This method does tend to thicken the join,

Fig. 5.282.

but this can be reworked and the correct wall thickness regained. There is always the chance that the tubing may deform at the mouth and will need reworking (reaming) to regain a round hole.

Alternatively the smaller diameter tubing could have its diameter increased by heating the end of the tube and flaring it out to meet the larger diameter using a carbon rod (Fig. 5.283) — the tip of which rests within the tube on the solid glass as the carbon rod is raised to form the flange on the molten glass.

This method will give a good control of the final diameter while the unequal join is made at the top of the shoulder, the smaller diameter also does not take as much heat to flare it out than does the carboning down method. A point to remember if dealing with delicate internal glass or when making an unequal join to a jacketed vessel.

Whichever method is used, one requires a rounded shoulder to finish off — heat the shoulder and rest a carbon flat on the smaller diameter while still heating the shoulder and blow which should produce a neat rounded join (Fig. 5.284).

Should a tapered join be required this can be made with or without the use of a carbon to shape the join by controlled blow and controlled

Fig. 5.283.

Fig. 5.284.

pull. It is always best to finish off with the flame to remove any tooling marks. Whatever the final shape of join the most important thing to look for is the wall thickness. A good consistent wall thickness is the standard to aim for.

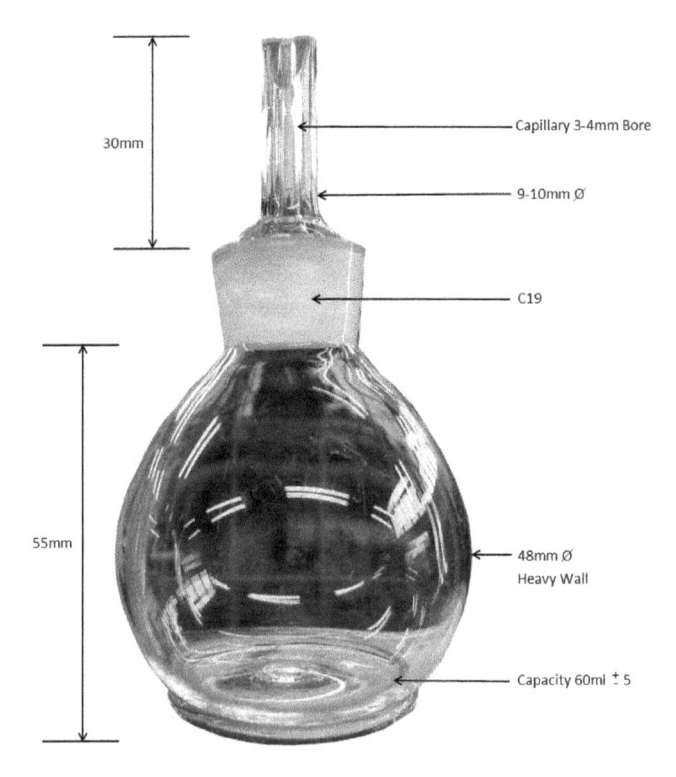

Fig. 5.285.

An Exercise

Specific gravity bottle (Fig. 5.285) — List of tools needed or desired:

Carbon graphite rods and plate, Vernier or digital calipers, 1:10 hexagonal carbon-tapered former, steel 1:10 grinding taper, brass or steel socket to grind the cone, carborundum powders grade 100 and 600, Linisher machine, optional single headstock grinding machine.

Consult the diagram and note that the volume is +/−5 mL, the capillary bore is 3–4 mm, the capillary ø is 9–10 mm. But the designation C19 is "critical" we must adhere to the sizes: large diameter = 18.8 mm, small diameter = 17.1 mm and ground length = 17 mm. If we fail to meet the required measurements the top and bottom of the cone and socket will not be in alignment.

We will have to form the cone oversize on the top and bottom diameters with same ground length then pre-grind the taper with the metal 1:10 taper and carborundum close to tolerance.

Form the socket undersize on the top and bottom diameters with the length to size then grind the taper using the steel tapered mandrel close to tolerance and finally marry both cone and socket together using the finer grade of carborundum.

The Socket and Body

I will describe the bench method using spear points, the stages are then readily transferable to the lathe.

Take a piece of 48 mm ø tube, pull two straight spear points leaving 100 mm between the shoulders, allow to cool, open one end and place the sealed end in the left hand. Heat the glass 30 mm from the right shoulder and use a "V" block to reduce the diameter, once the constriction is close to 20 mm diameter move to the left and blow the shoulder into a gradual arc from what will be the small end of the taper. (There will be a mark formed by the edge of the "V" block.) Heat the remainder of the glass to gradually form a taper to the right of the V block mark, measure 20 mm to the right of the mark and pull off the spearpoint and "blow out" the tubing, we should now have a partially formed body on spearpoints with a rough open taper (Fig. 5.286).

Fig. 5.286.

Fig. 5.287.

Take the 1:10 hexagonal taper, we are going to mark the length of the taper using an ordinary pencil, you will by now have noticed that whatever the size of sockets A,B,C,D the large diameter is the same size, i.e. 18.8 mm (finished ground size). Take an ordinary B19 socket put the carbon former inside and mark the former where the top of the socket lies (Fig. 5.287).

The ground length of the C joint we are about to make is 17 mm, measure this down the taper and mark, as this is the finished size we need to make allowance for the grinding, so make a second mark 1–2 mm further down the taper, with practice the grinding depth can be reduced but not yet!

On the taper we now have a mark at the narrow end of the socket with the indicated grinding depth. Try the carbon taper in the rough formed socket to reassure oneself that the taper is not too wide.

Heat the taper in a gas air oxygen flame as the glass softens, gently place the carbon taper inside, keep the carbon static and in line while you rotate the glass with the left hand, one will see and feel the glass forming to the taper. Do not push too hard as ripples in the glass may form rendering the socket unusable.

Fig. 5.288.

Watch the bottom line on the carbon align with the V block mark on the body, once there remove the carbon, allow the glass to cool a little and retry the carbon to check the mark. If you have not managed to keep the carbon aligned centrally there will be an increase in size at the large end of the socket — one will feel the gap as the carbon will not be in contact along the full length of the taper — heat the opening, allow it to reduce in size a little and re-establish the taper and allow to cool (Fig. 5.288).

At this point take the metal grinding taper, place in a normal socket and mark the top onto the taper — measure the total length of bottle from this point on if the taper will be greater than the length of the body of 55 mm it will impinge on the base whilst grinding. Therefore one will need to grind the taper and complete the body later.

If all is well place a B19 cone in the socket using corrugated carbon tape between the two surfaces, seal the end of the B19 with a cork and place in the left hand, open the spear point and shape the body using the diagram and callipers to confirm diameter and length, pull off the spear point and flatten the base against a carbon plate, finally anneal the glass (Fig. 5.289).

Fig. 5.289.

Take a 30 cm length of capillary tubing of the size indicated, make sure that the cut ends are flat without any sharp spikes. Seal one end and place in the left hand, mark the centre and push up a large button, then a second and third while moving the flame to the right, occasionally blowing to maintain the bore, although the shoulder is gradual push together to increase the diameter to the left of the gathered buttons until a diameter of 20 mm is reached — then melt and shape the taper using the callipers frequently to check diameters and length remembering to allow for grinding. Once satisfied anneal fully. Remove the glassware from the annealing oven and measure the length of the socket, flatten and straighten the rim using the Linisher (Fig. 5.290). Take the metal 1:10 grinding taper and fit this into the single headstock grinding machine, moisten the taper and smear with carborundum slurry, start the machine slowly and gradually put the socket onto the tapered shaft to feel the grinding of the taper, be careful not to press too hard as the glass might "catch" and stick, move the glass off and on the taper until it can be felt to be grinding evenly to produce an even matt surface.

Stop, remove the glass and wash off the carborundum to check the measurements, repeat the process until within 1mm of the final diameters — wash

Fig. 5.290.

Fig. 5.291.

and dry the glass and put to one side for a moment, remove the metal taper and put in its place the shaft of the capillary (Fig. 5.291).

In preparation for the grinding of the cone we either use proprietary grinding tapers or make one of our own.

Fig. 5.292.

Take a sheet of brass approximately 40 mm × 100 mm × 1.5 mm thick and form it around the metal 1:10 taper to form a tapered tube with two "wings", grip the wings in a vice and squeeze to confirm the taper, a blow with a wooden mallet at the large end of the former will stretch the brass to ensure that the 1:10 profile is correct (check it against a standard glass joint) (Fig. 5.292).

Apply carborundum slurry inside the brass taper and use this to grind the cone, confirm the diameters constantly. One now has two items of glass with corresponding tapered joints which will now need marrying together. Wash off all previous grinding paste. Use fine grade carborundum to complete the task until both the top and bottom of the joints align.

Cut the capillary stem to length, using the Linisher grind two parallel facets in the capillary and finally glaze the two faces after warming carefully — hold the C19 cone in a spare B19 socket — after glazing, anneal (Fig. 5.293).

The use of diamond impregnated tools to grind cones and sockets is being used more and more as it is a far cleaner method than using carborundum powders. The initial cost of this would probably be more suited to larger scale production rather than the occasional "one off" laboratory glassblowing situation.

Fig. 5.293.

5.19 Flange Making — Dewar Seals, Sinters

Sinters (filters) are as follows:

Specific size glass frits semi fused into discs to create various sized porosity filters, 0 being the coarsest and 5 the finest.

Their Application

Porosity 0 used for gas distribution in liquids with low gas pressure and the filtering of the coarsest precipitates.

- Porosity 1 — liquid distribution, extraction apparatus for coarse grained material, support for loose filters when using gelatinous precipitates.
- Porosity 2 — preparative fine filtration with crystalline precipitates and Mercury filtration.
- Porosity 3 — analytical filtration with medium-fine precipitates. Filtration in cellulose chemistry.
- Porosity 4 — analytical work with very fine precipitates.
- Porosity 5 — ultra fine filtration.

The Exercise

Select a 20 or 25 mm diameter disc of either porosity 0 or 1 and a 30 cm length of tubing into which the disc will slide. Attach a blowing tube and swivel to the tube and place in the left hand chuck of a lathe, slide the disc into the tube (Fig. 5.294). Take a 15 cm length of 9 mm ø tube, seal one end and place in the right hand chuck, move the right hand chuck so that the edge of the 9 mm ø tube is positioned 25 mm within the larger tube. Start the lathe, the disc will move within the tube. Blow hard and the disc should move rapidly towards the inner tube against which it should rest and be held vertically so long as a continuous pressure is maintained (Fig. 5.295). Use a hand torch to warm the area to be sealed. Once the sodium flare is seen, reduce the size of the flame and aim it in line with the disc edge. As the glass softens, use the edge of a carbon/graphite plate to press the glass onto the sinter edge. Remove the support tube by moving the right hand chuck away while you soften the flame a little and heat the edge of the open tube. Use the carbon flat to reduce its diameter to approximately 9 mm (Fig. 5.296). Move the right hand chuck to the left, so returning the support tube — confirm that the two diameters are similar in size — heat both edges and bring together to make an unequal join, heat the initial join then the shoulder, alternate between heating and blowing the shoulder to finish with an

Fig. 5.294.

Fig. 5.295.

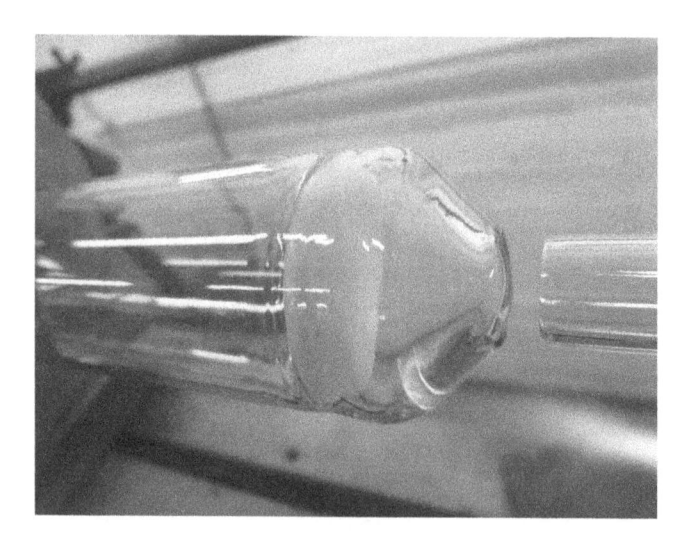

Fig. 5.296.

even rounded join. Warm off in the flame then place immediately in a hot oven and anneal.

Note: The seal is more of a physical join not a fused or run in seal as too much heat will "blind" the sinter as the sinter face will glaze over. Look closely at a

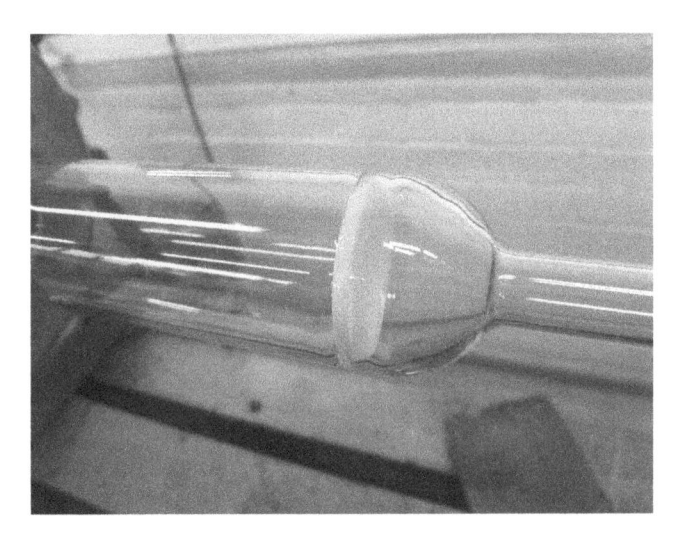

Fig. 5.297.

porosity 0 sinter after the seal is made, on the edge of the seal only the high spots of the sinter make contact with the inner wall of the tube (Fig. 5.297).

Specific diameter sinters can be made by simply drilling a larger disc with a diamond impregnated core drill — the disc would then need washing and drying prior to sealing. Alternatively, a disc can be resized by rotating against a Linisher belt (Fig. 5.298) — examples of suitable holders are shown (Fig. 5.299).

Larger diameter Sinters have a domed radius, the design of which gives strength to the filter when a negative pressure is used to draw the filtrate through the filter.

The maximum pressure to be exerted on a sintered disc is one atmosphere.

Once sealed and annealed the sinter can be used in a column as a packing support. Sinters can be sealed within a jacket when the chemistry demands temperature control.

Dewar Seal

This type of seal was developed to create an annular space which would then be silvered, evacuated and sealed off to make a vessel used throughout the world to contain hot or cold liquids, over the years the Dewar seal has been

240 *Laboratory Scientific Glassblowing: A Practical Training Method*

Fig. 5.298.

Fig. 5.299.

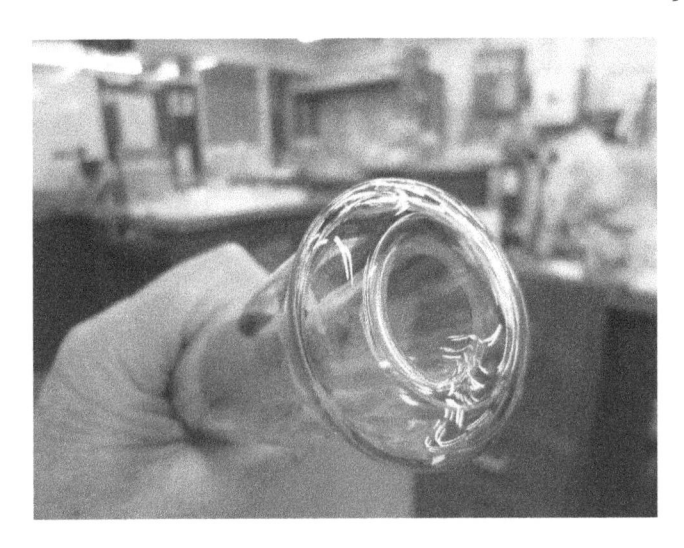

Fig. 5.300.

found to be a flexible and useful type of seal for jacketed columns where both inner and outer tubes are brought together in such a way that there will be little or no stress as the glass flows together in a domed seal with the whole seal acting like a flexible bellows (Fig. 5.300).

The inner will be sealed and rounded off when making a Dewar vessel and opened when making a column — an interesting point to note: when blowing the seal, there is the possibility of forgetting to temporarily seal the inner column tube.

Supporting the inner centrally within the outer needs to be considered carefully especially when making a Dewar vessel with an internal dome. One more problem prior to starting a vessel is the question: Will the end of the outer tube be rounded off prior to the seal? or, Is the annular space sufficient to allow the outer tube to be drawn down after the seal has been made and any packing removed?

The configuration of one type of cold trap is made using the same method as a standard Dewar vessel but with a greater annular space, side arm and bottom run off so that the vessel can be accommodated within a vacuum line; the system itself creates the evacuated annular space.

When used to make a jacketed column, the Dewar seal at the top of the opposite end will most likely have an internal seal with little or no

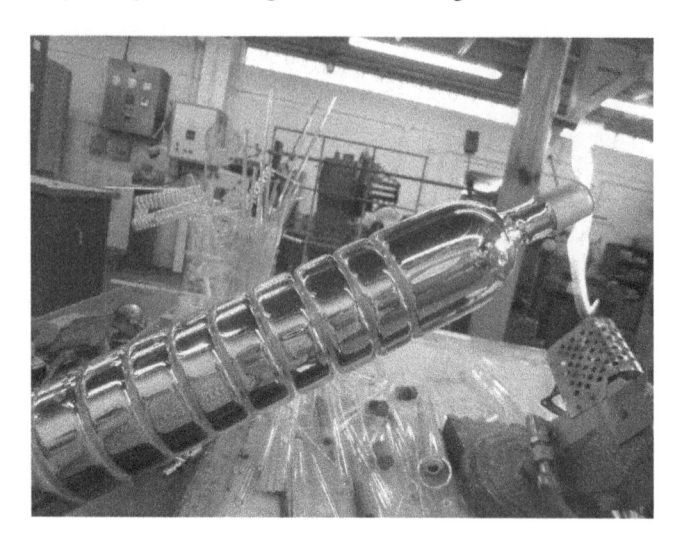

Fig. 5.301.

movement for expansion when temperature changes occur, it is advisable that some form of bellows are incorporated in the outer wall of the column to the rigid bottom seal to prevent it from pulling the entire column apart (Fig. 5.301).

In recent years I have found that an internal expanding holder to be the best means of support for the inner tube as one can be sure that the internal tube will be held centrally especially when making vessels of larger diameters (Fig. 5.302). The one thing to remember when using this type of aid is to pre-warm the metal as its expansion rate is far greater than the glass. If one fails to warm the metal, it will split and break the glass which can be quite annoying.

The Exercise

Take a piece of 26 mm ø tube 40 cm long, mark the centre, pull a spear point, burn off the spear point and blow a domed end. Take a piece of 46mm ø tube 30cm long to support the domed tube within the larger tube using either corrugated paper or metal mesh with 10–12 mm protruding past the end of the larger tube. The protrusion should coincide with the distance between the internal diameter of the inner and the outside diameter of the larger tube. Seal the larger tube with a cork, tube swivel and blowing

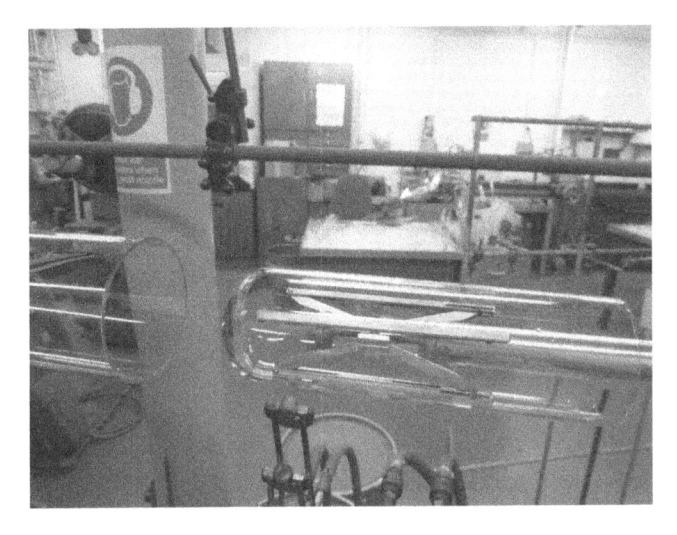

Fig. 5.302.

Fig. 5.303.

tube. Place in the left hand chuck of a lathe. Start the lathe to make sure that the inner tube is lying centrally. If not central, either repack the inner or use the tip of the carbon paddle to touch the inner tube as the lathe turns slowly which may well bring the inner to the centre (Fig. 5.303). If this does not work — repack; do not try to make the seal if the centre tube is offset.

Fig. 5.304.

A hand torch should provide sufficient heat for the diameters involved. Start the lathe once more. The flame should play predominantly on the inner tube which when starting to soften is brought vertically with a carbon rod or thin plate to make contact with the outer where the two are "run in" — withdraw the flame while blowing so as to stretch and remove the initial join — return the (Fig. 5.304) flame and repeat heating the seal concentrating on the outer rim — I always tend to over blow the outer while being careful not to overheat the inner which may constrict — if this happens regain the inner diameter by reheating and reaming out with a carbon while blowing once more. When making a column, the longer lengths involved must be central — ensure that the packing is adequate and the inner secure.

"Cup sockets" with an extended rim are designed for such an operation where the seal can be made without distorting the joint (Fig. 5.305).

Flat Flanges

Flat flanges and wide buttresses give good access to reaction flasks, they can accommodate a variety of joints from the central stirrer to thermocouple pockets, condensers, inlets for solids and liquids and outlets for vapours all without restricting access to the flask (Fig. 5.306). Flat flanges and buttresses are press moulded as lids and closures to which joints may be added or joined

Fig. 5.305.

Fig. 5.306.

to a short section of tube to enable fusing to a reactor body or column. Often, one is called upon to make equipment with non-standard flange fittings or to have apparatus fit directly to standard size flanges. To make flanges on a lathe there will need to be fitted a tooling bar. That is a horizontal bar on which a carbon plate can be fixed at a right angle to the work.

Fig. 5.307.

I find it easier if the carbon lies flat on the glass from the start with sufficient glass to make the flange protruding to the right of the carbon edge. The carbon should be substantial and at least as thick as the depth of the intended flange (Fig. 5.307).

Throughout the heating process it is best to keep the flame from playing directly on the carbon as it will deteriorate so much so that it will be worthless in a short period of time.

How much should the tubing protrude?

Big question! If the flange is to be 30 mm deep and we use this as a measurement, the resultant flange will be about 20 mm deep and very thin — useless! We have to realise that when shaping the flange against the carbon it has to bend around the carbon to make a 90° vertical bend all of which uses glass, so we have to allow for the thickness of the eventual flange, we need a sharp corner from the flange and not a radiused bend to the flange — we have to have sufficient thickness of flange so that after final grinding is carried out we have the correct final thickness. I am aware of a calculation for the amount of glass required using a method which involved displacing a volume of liquid all of which is very nice and theoretical but not much use to a glassblower. There are many variables: wall thickness of the tubing from which the flange is to be made, width of flange, final flange thickness and the technique used to create the flange, not to mention the experience of the glassblower.

The Exercise

As a rule of thumb when using medium wall tubing: protrude the glass from the front face of the carbon by ⅓ of its diameter. Ensure that the cut edge is straight, use the burner to glaze the face of the tubing so that when a flat carbon paddle comes into contact with the glass it slides against the glass rather than a sharp edge which may cut into the carbon. After glazing move the flame in line with the edge of the carbon former, heat at this point while using a flat carbon paddle to thicken the glass by gently pushing the glass towards the carbon former — one should see that the glass is thickening just in front of the carbon former, continue until all of the tubing has become part of the "thickening", continue to heat vigorously to produce an even mass of glass.

Use the carbon paddle to form the flange by inserting it into the tube, switch off the flame and lift the glass vertically while pushing it against the carbon former. Once almost vertical, withdraw the carbon paddle and press the paddle against the front face of the flange — press hard — whilst still hot, check that the inner tube has not constricted by placing the paddle inside the tube and hold horizontally. Now press the flange against the carbon former until the glass cools (Fig. 5.308).

Fig. 5.308.

Fig. 5.309.

The aim is to form a shape that is slightly thicker at the hole tapering out to the edge of the flange, which in itself should be thicker than the final specification. This will ensure that the flange to hole will be crisp and at 90º after grinding. One may use a round carbon rod to flange out the glass if preferred. A rod will allow the glass to roll over its surface giving a smoother finish to the flange face (Fig. 5.309).
Note: Make a note of the tubing used to successfully produce a given diameter flange for future reference. The flange must be annealed before grinding.

Use diamond impregnated discs of increasing smoothness after using the coarse grades to take the front face of the flange close to the final thickness.

Carborundum powders mixed with water can be used on a rotating lapping machine, which consists of a rotating cast iron disc onto which the carborundum is spread. Wash both the disc and the glass between each grade of carborundum until an even matt surface is reached (Fig. 5.310).

The same final finish can be reached using carborundum powders and a piece of plate glass, again washing between each grade, the flange is moved in a figure of eight to distribute the carborundum and the wear on the plate. The final surface is adequate but the time taken tedious.

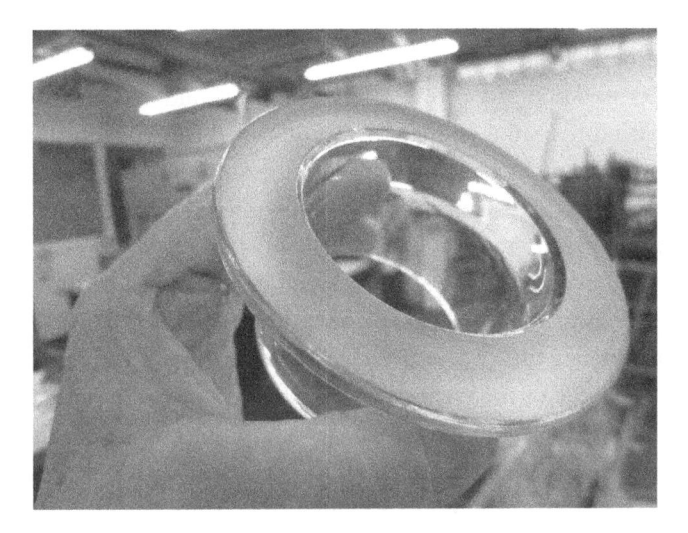

Fig. 5.310.

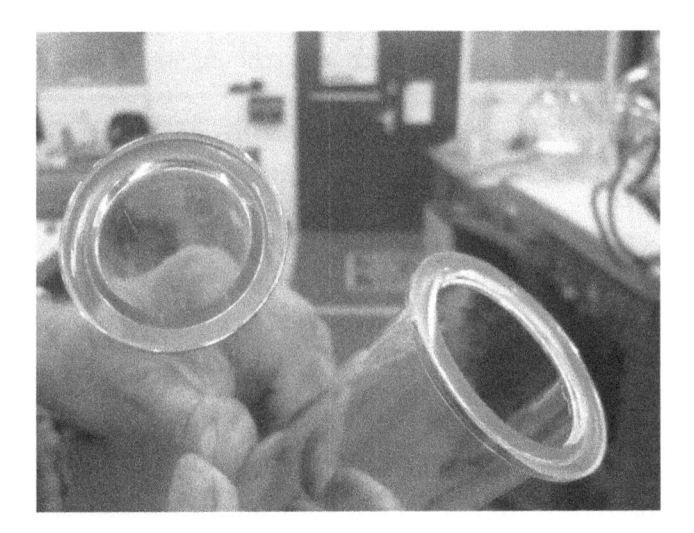

Fig. 5.311.

Flanges come in all sizes and specific profiles, for example small square minute profiles may be needed to fit existing equipment (Fig. 5.311).

A profile can be cut into a carbon former and the hot glass encouraged into shape with a paddle or rod.

5.20 Safe Working Pressures (Positive) — Vacuum

When used under positive or negative pressure and especially when also working with differential temperatures, additional care must be taken. Glass apparatus that is under pressure or vacuum should only be subjected to further stress with extreme caution, as the individual resulting strains are additive and could readily result in failure.

To avoid stresses in the glass, evacuated vessels or vessels under pressure should not be heated on one side or heated with an open flame.

When working under pressure the maximum figures indicated in manufacturer's catalogue or the enclosed graph should not be exceeded.

Before using glass equipment under vacuum or pressure, it must always be visually inspected to check that it is in perfect condition. Damaged glassware should not be used for work under pressure or vacuum for safety reasons.

Never subject glassware to sudden pressure changes e.g. always repressurise evacuated glass apparatus slowly.

Laboratory glassware with a flat bottom should not be used under pressure or vacuum.

The plastic coating of laboratory bottles has no influence on pressure resistance.

The graph (Fig. 5.312) shows the relationship between the wall thickness and outside diameter to the maximum working pressure in KN/M^2 = kilo newton per square inch (P.S.I.).

The maximum working pressure applies at 25°C and assumes that the glass is free from scratches, chips or cracks.

The graph only applies to glass which is freely supported and not subject to clamping forces.

The graph only applies to fully annealed "tubing" that has not been worked.

That is, without any form of side arms, internal or external seals, coils, flat bottoms etc.

In other words once an article of whatever size or shape has been scientifically glass blown, the figures are no longer valid!

The graph is a guide only as to the strength of the material prior to working but is a good indication as to the relative strength of the glass based upon its wall thickness.

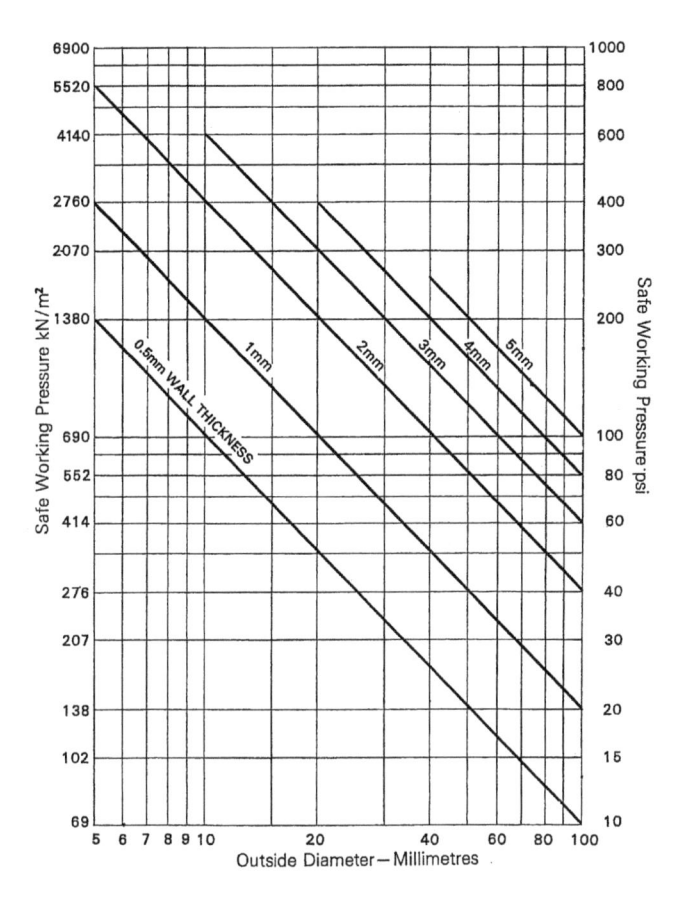

Fig. 5.312.

When we, for example, discuss with the customer their requirements for a simple vial 25 mm diameter to withstand 40 P.S.I. or 276 KN/M² a 25 mm ø tube with a wall thickness of 0.5 mm will intersect at 40 P.S.I. but a piece of tubing 25 mm ø with a 1.0 mm wall thickness will have a safe working pressure of: 80 P.S.I. or 552 KN/M² and that 25 mm ø tubing with a 2.0 mm wall thickness will have a safe working pressure of 170 P.S.I. or 1,1072 KN/M².

My advice would be to make the "simple vial" in tubing with a 2.0 mm wall thickness in the knowledge that any work done on the "simple vial" that I would have to ensure that the wall thickness throughout would have to be 2.0 mm or better so that the glassware could be used well within the indicated safe working pressure.

The graph is a quick guide only. Err on the side of caution if the potential pressure within a vessel is in anyway close to the indicated safe working pressure advice against its use.

Point out that glass would be unsuitable so much so that a metal vessel would be more appropriate.

Note: Heavy walled large vessels for use in industrial chemical plants have a maximum safe working pressure set at two Bar or two atmospheres, to a degree it is because of the environment to which the glassware is exposed.

The environmental conditions in which glass is used and its effect on glass can be shown in the case of a borosilicate boiler sight glass (Fig. 5.313) which came into contact only with steam and boiling water.

The diameter is 18.5 mm with a wall thickness of 3.0 mm which according to the graph will give a safe working pressure of 320 P.S.I. pounds per square inch 2.200 KN/M^2 kilo newtons per square metre.

The boiler sight glass was in place for 12 months after which time the wall thickness had diminished. It had been dissolved by steam and water (Fig. 5.314).

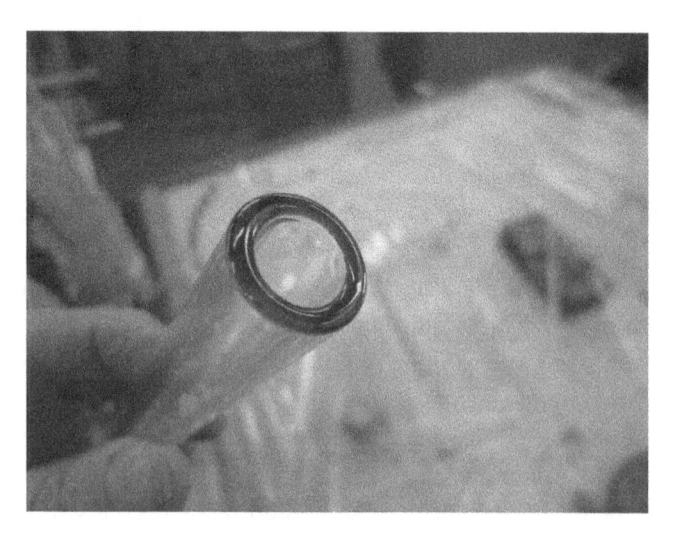

Fig. 5.313.

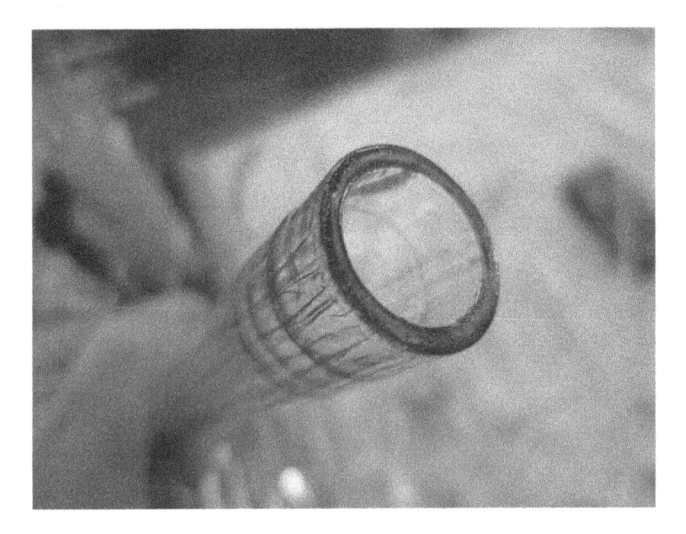

Fig. 5.314.

The overall diameter had remained at 18.5 mm but the wall thickness was 1.9 mm which gave a safe working pressure of 215 P.S.I. pounds per square inch 1482 KN/M^2 kilo newtons per square metre.

All glassware used under pressure or vacuum should be checked regularly for signs of chemical attack, scratch marks or cracks as a matter of course even to the point of instituting a formal inspection at set periods of time.

If it is felt that if there is the slightest possibility of glassware under pressure having a sudden or uncontrollable rise in pressure there should be in place a means of controlling the situation with either a bursting disc within the system which will vent safely to atmosphere or have the entire glassware contained within a "blast bay" which is a substantial purpose built area equipped with alarms, bunds to contain spillage and equipment to deal with chemical fires where all the safety controls are external to the bay. Glass under vacuum must be used under control condition where full training is given to those who are expected to use and service the equipment — shroud the glassware whenever possible with open weave plastic which will help to contain any shards of glass because if there is a failure under vacuum, the glass will first implode then reverse its direction of travel and inevitably cause injury.

Fig. 5.315.

The open weave plastic (Fig. 5.315) is available in vivid colours, which mean that when in use it raises the awareness of all present within the laboratory that vacuum is being used, it raises the overall awareness of the danger in addition to the regulatory signage required.

5.21 Glass to Metal Seals, Recrystallised Alumina to Glass Seals Graded Seals

The key to glass to metal seals is the "coefficient of expansion" of both the metal and glass. When hot they might have a similar expansion rate but diverge at or near the transition point, the result being separation of the two materials.

This can be overcome in a limited number of ways by selecting a glass similar to a metals expansion rate, for example platinum in soda-lime glass, tungsten in borosilicate with perhaps an intermediate sealing glass designed specifically to accommodate the difference in expansion.

Tungsten, nickel–iron–cobalt and nickel–iron–chromium metals for example use an intermediary oxide layer to make the glass–metal bond.

Where the expansion rates are vastly dissimilar, the unmatched seals such as copper and glass use the ductility of the metal combined with a thin feather-like shaped edge to make the join.

Stainless steel to glass also combines a thin contact edge and an oxide layer to make the seal.

Platinum in soda-lime glass is a useful means of making a vacuum tight seal where an electrical connection has be made through a glass wall, with the *proviso* that this is best used at the experimental stage in a glassblowing workshop attached to a research establishment rather than a production situation which tends to be far more automated.

I have on occasion made an entire vessel in soda-lime glass complete with platinum electrodes where the temperature in use has been stable, thus overcoming the need for graded seals and the like.

You will find that soda lime glass wets and "sticks" to platinum very easily.

The Exercise

Take a piece of platinum wire approximately 0.5 mm diameter and 25 mm long.

Hold the wire in a pin vice after having drawn some soda-lime glass rod down to 2 mm diameter, heat the rod in the tip of a small gas and air flame. The glass will readily melt, touch the tip of the rod onto a wire and rotate the pin vice as the molten glass wraps around the wire, burn off the excess rod and continue to rotate the bead of glass in the top of the flame to make a nice even spherical bead (Fig. 5.316).

The platinum should be seen through the glass to be bright clean and without bubbles. Remove the platinum from the pin vice and place on a warm surface directly underneath the flame to keep it warm for a few moments.

Take a length of soda-lime glass 10 mm ø and 20 cm long that has also been resting underneath the flame to be pre-warmed. Introduce the end of the tube to the hot air zone which is beyond the tip of the gas/air flame, lower the glass into the top of the flame and as the glass softens use tweezers to draw off the tubing, remove the spear point and blow a rounded end.

For those not used to soda glass, be aware that it is a much more gentle form of glassblowing by comparison with the harder borosilicate.

Reduce the size of the flame and aim at the very end of the bulb, withdraw and touch the end with a piece of rod then draw a small point, then either burn off and blow a small hole or mark and tap off with the knife — use the tweezers to grip the platinum and try the bead in the hole to see if

Fig. 5.316.

the bead will sit on the rim of the hole — if not, adjust with a carbon reamer to increase or use the flame to decrease the size.

When satisfied, hold the platinum and bead in place as you return to the tip of the flame, as the glass softens, push the bead gently to make contact with the rim — remove the tweezers while using the flame to "run in" the seal and dome while inclining the glass.

Blow to retain the dome shape. One will be surprised at the speed at which the melting stage happens (Fig. 5.317). Move the glass into the hot air zone above the flame tip, reduce the amount of air in the flame to make a luminous flame, continue to "warm off" the glass then stand the glass upright while the glass cools to room temperature.

Once cool, wrap the wire around a 3 mm diameter rod to form a loop which can be gripped to make an electrical contact (Fig. 5.318).

The flame annealing process has to be more rigorous for soda-lime glass than for borosilicate, I would also pre-warm both the tools, the bench and the glass with the radiated heat from a luminous flame before starting work with soda-lime glass.

Second Exercise

Repeat the first exercise until competence and confidence has grown along with an understanding of how the glass reacts.

Fig. 5.317.

Fig. 5.318.

Prepare a platinum wire and bead. Dome one end of a length of glass tube 10 mm ø 20 cm long.

Whilst the glass is still hot, pull or blow a hole 25 mm from the top of the dome, prepare and rest the bead on the rim of the hole. Using a fine gas/air flame run the side seal in, I find that aiming the flame directly at the

Fig. 5.319.

platinum bead produces a well run in seal as the platinum itself glows and heats the seal from within, while occasionally giving a small controlled blow (Fig. 5.319).

Use the tweezers to align the platinum centrally before annealing in the hot air zone. Once cool, wrap the wire around a 3 mm diameter rod to form a loop which can be used to make an electrical contact (Fig. 5.320).

Soda lime glass can be difficult to re-warm especially after being previously worked. The glass is quite robust but susceptible to thermal shock. It can be quite satisfying to have completed a task with this type of glass.

Tungsten to Borosilicate

This glass to metal seal is made via the tungsten oxide layer, the preparation of which will determine the quality of the seal.

Tungsten rod and wire are made by one of the two methods.

The better alternative is to use swaged tungsten. Swaging is a process of hammering to shape the wire or rod which is available in sizes up to 2.5 mm diameter. The material made in this way is quite dense and without the striations caused by the alternate process, where the tungsten is drawn through a die. When this form of wire is sealed through glass, the striations or die scratches may cause leakage due to uneven surface sealing.

Fig. 5.320.

There would therefore be a need to seal the end of the wire with a bead of nickel.

The tungsten must be cleaned thoroughly before the surface is oxidised prior to sealing.

One method is to rub the hot tungsten with potassium or sodium nitrite.

I much prefer to use electrolysis to ensure a clean surface on which to oxidise the metal.

Equipment required:
Beaker 600 mL
Distilled or de-ionised water
Sodium Hydroxide
Nickle sheet 30 mm × 120 mm
Small 9 volt battery
Crocodile clips
P.P.E eye protection
Gloves

Method:

Make a 10–15% solution of sodium hydroxide.

Fig. 5.321.

Fig. 5.322.

Connect the negative side of the battery ⊖ to the tungsten and the "positive to the plate" (Fig. 5.321). Immerse the nickel plate in the solution followed by the tungsten and watch as bubbles emanate from the tungsten surface which becomes a slate grey in colour (Fig. 5.322), disconnect the battery and wash the tungsten, keep immersed in distilled water until required.

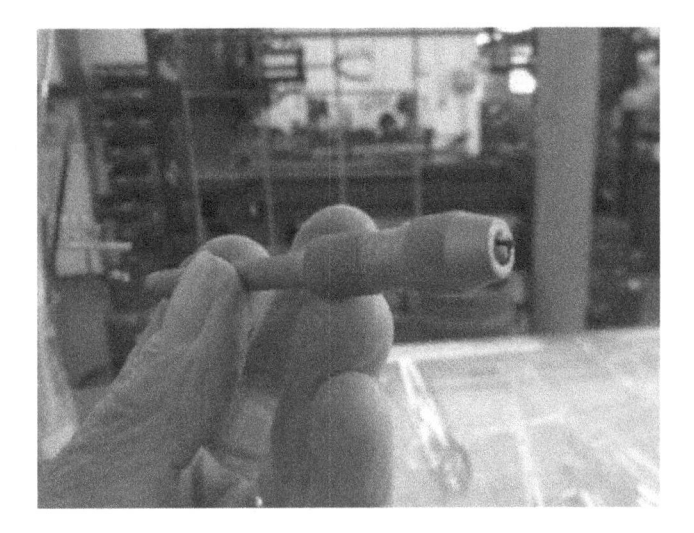

Fig. 5.323.

Do not handle the clean tungsten, when about to make the seal, hold the tungsten in a pin vice (Fig. 5.323).

Pull a series of spear points, select and cut a piece 20–25 mm in length that will just slide over the tungsten, this will be the initial seal, select two or three slightly shorter lengths of tubing with increasing internal diameters, these will shroud the initial sleeve before a final bead is made with which to make the seal.

The Exercise

Hold the pin vice and tungsten in the left hand, have the various size sleeves close to the right of the burner together with a pair of tweezers.

Heat the tungsten evenly moving through a gas/oxygen flame smartly from left to right, repeat three times, after the third pass slide the initial sleeve over the tungsten, soften the flame with air, rotate the glass in the very top of the flame, slowly pass the glass once more from left to right while watching closely as the glass shrinks and seals to the tungsten surface (Fig. 5.324).

The process depends on you expelling the air as the glass moves from left to right.

Time the movement as the oxidised layer contacts the glass. As soon as you reach the end, slide a shroud over the initial seal, repeat the process of

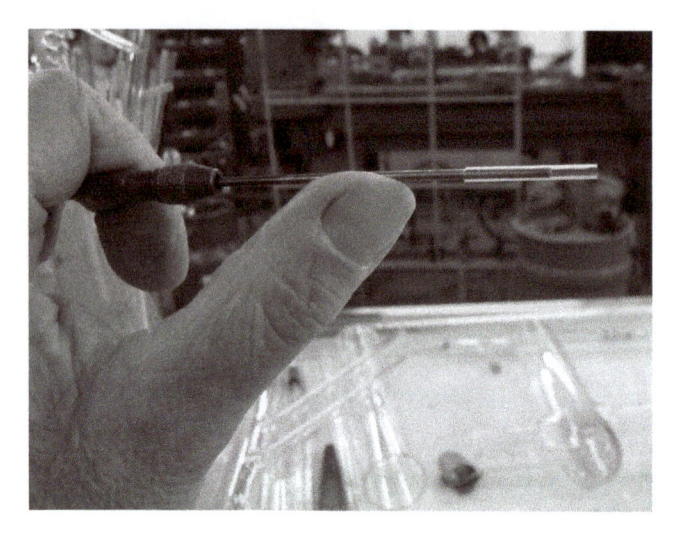

262 *Laboratory Scientific Glassblowing: A Practical Training Method*

Fig. 5.324.

Fig. 5.325.

the sealing gradually from left to right, you have the option of repeating with a further one or two shrouds.

When completed, look carefully at the colour of the seal which should range from deep gold to a mid-brown (Fig. 5.325).

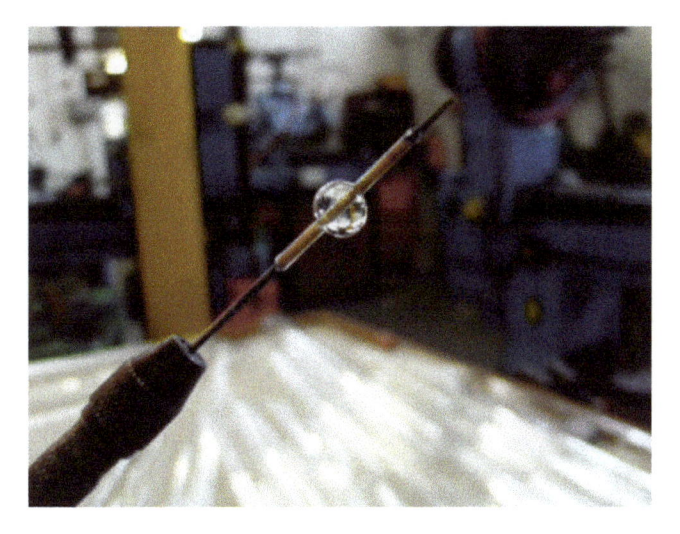

Fig. 5.326.

If the seal is black, the oxide layer has been too thick and has burnt, if light in colour "flash" the seal through a hot oxygen/gas flame quickly which should improve the colour and the quality of the seal.

Make a bead around the centre of the seal and melt in place (Fig. 5.326).

Flame anneal and allow to cool, if working from a long length of tungsten cut to length using a carborundum grind stone.

The prepared seals can be used immediately or annealed, pre-warmed and sealed into a glass envelope as per platinum seal, longer length of tungsten acting as posts or supports will require supporting internally as the final seal is made to the glass tube or envelope.

There are intermediary sealing glasses specifically for use with tungsten which have an expansion rate compatible with the metal.

All of which are application and cost dependent (Fig. 5.327).

Nickel–iron–cobalt or chromium alloys are being used more and more due to their robustness and vacuum integrity.

The preparation of the metal requiring its controlled heat treatment, cleaning and profiling prior to sealing lies well within the compass of glass engineering and beyond the compass of most standard glassblowing workshops (Fig. 5.328).

The metal that is in contact with the glass has to be profiled to give a radius edge without any sharp edges.

264 *Laboratory Scientific Glassblowing: A Practical Training Method*

Fig. 5.327.

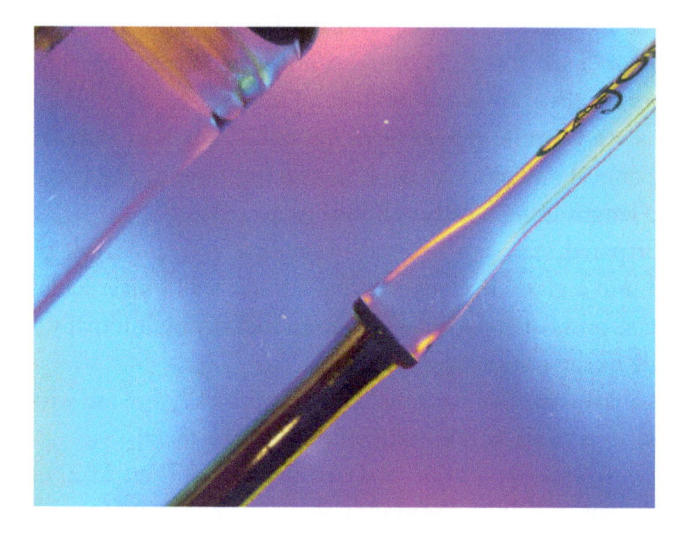

Fig. 5.328.

Background:

Nickel–iron–cobalt alloys were first developed in the 1930s and were referred to as "Fernico", after the original patents expired the glass was called variously as Nilo K, Dilver P, Vacon, Kovar.

The compositions were all similar, for example, Nilo K is Nickel 29.5%, iron 53% and Cobalt 17%.

There are a number of variations of the alloy with the addition of chromium to make the metal compatible with soft lead and soft soda-lime glass.

The metal has to be prepared by annealing within an atmosphere of cracked ammonia and hydrogen between 850°C and 1000°C. Decarburisation to prepare for sealing is carried out in an atmosphere of wet hydrogen at 900°C–1050°C for 1 hour.

To create the metal oxide layer, the metal can be heated in air between 600°C and 1000°C depending on the thickness of the oxide film required.

Degreasing is followed by boiling in 10% sodium hydroxide for 5 minutes then heated for a further 5 minutes at 80°C in a 10% hydrochloric acid solution.

The seal area is then grit blasted followed by ultrasonic cleaning with alcohol!

The method of sealing to rod or wire is similar to that of tungsten where sheaths of tubing are slid over the pre-oxidised wire and heated from one end to exclude the air as the flame progresses along to complete the seal, a second or third sheath is added and a final bead with which to make the button seal.

The tubular seals start with a close fitting tube, the metal oxidised to show a dull grey surface, the glass tube slid over the metal and using a carbon flat paddle pressed onto the metal, the excess removed and the glass rolled over the edge and inside the tube a second sealing glass is joined, the excess removed and the final join made to the desired borosilicate glass as in a straight join.

The seal is annealed and finally the oxide layer is removed by either acid pickling acid bright dipping or electrolysis.

Copper to Glass

Housekeeper unmatched seal was developed in the 1920s. Tubular seals between copper and soda-lime glass depend on the profile of the copper having a thin edge which is spun or rolled to a thickness of 0.25–0.1 mm (Fig. 5.329). The glass is sealed to an oxide layer, the difference in expansion rates is vast but the ductility of the copper bridges the gap.

$$\text{Copper} = 17 \times 10^{-6} \text{ per }°C$$
$$\text{Soda-lime glass} = 9.6 \times 10^{-6}$$

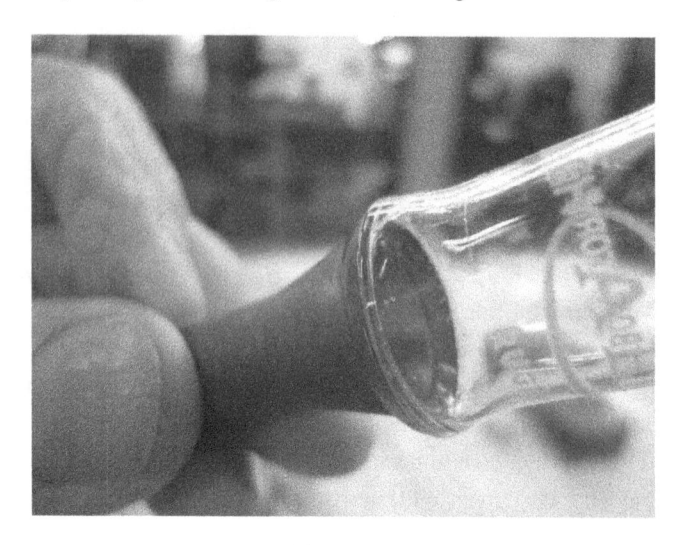

Fig. 5.329.

Copper clad nickel iron wire has a nickel iron core with 43% nickel 57% iron with a copper sheath.

This material is used for crimp seals which can be found on incandescent light bulbs — the maximum diameter of the wire must be 0.7 mm so as to avoid cracking as there are no options for the wire to deform as in the tubular Housekeeper seal.

Stainless steel uses the thin profile Housekeeper design. Non-magnetic stainless steel is used and prepared by heating in dry hydrogen at 1065°C for 15 minutes, then machined to give a sealing edge of 0.025–0.038 mm.

There are very specialised methods of making glass to metal seals one of which is to use high frequency induction heating.

The heat source being a high frequency electrical field, any glass in contact with the metal softens and flows over the metal to make the seal usually in an inert atmosphere. The advantage being that the control over the work is both greater and cleaner than with a flame.

I would advise that understanding of the following engineering techniques be sought:

Argon arc T.I.G welding

Brazing — silver soldering

Gas welding

Friction welding

Induction brazing

Vacuum brazing
Hydrogen brazing
Electron beam welding
Metal arc welding
Soldering.

Recrystallised Alumina to Glass Seal

Alumina–Aluminium Oxide (Al_2O_3) uses = Thermocouple protection tubes, crucibles and insulators.

The material that we will use to make the alumina to glass seal is a material combining Alumina and Silica called Mullite 361.

Synthetic Mullite $Al_2O_3\_SiO_2$ is available in both porous and non-porous forms. It combines high strength with good thermal shock resistance which will operate at 1600°C and is therefore used in furnace tubes.

The material to use has an alumina content of 60%.

When used as a furnace tube a suitably size socket making a direct connection to the mullite enables the passage of an inert gas or wet hydrogen through the furnace to be a practicable option.

Whether using an intermediary glass or a direct borosilicate seal, the initial warming of the alumina and the final shape of the seal is of great importance.

The operation is best carried out using a lathe with the alumina set in the right chuck and the required joint set in the left chuck attached to a blowing tube via a swivel.

The borosilicate should be at least ⅓ greater in diameter than the alumina.

The Exercise

Set a clear feather-like flame close to the edge of the recrystallised alumina, but not directly on the alumina which needs to be prewarmed in the hot air zone for 2 minutes after which the very edge of the alumina may be brought into contact with the flame.

On the assumption that the alumina does not crack split or suddenly fall apart, continue to warm the material within the feather-like flame for a further 2 minutes.

There will be some sodium flare to indicate that the material has reached a sufficiently high temperature to start the sealing process.

Once the Alumina has been warmed thoroughly, increase the amount of oxygen in the flame which should now accept the heat. Take a piece of rod 5–6 mm ø using a slow lathe speed heat and trail the rod around the alumina 10–15 mm from the tube edge, burn off the surplus rod and heat vigorously until both glass and alumina are "glowing" hot.

Use a flat carbon paddle to press the glass onto the alumina to produce a flat ribbon around the alumina.

Bring the edge of the borosilicate tube in line with the right edge of the ribbon of glass, heat the edge of the tube and using the carbon plate very gently bring the tubing into contact with the centre of the ribbon.

Once contact has been made, blow to confirm the seal, then angle the flame at 45° to the seal and shoulder, heat once more and blow to create a rounded shoulder so that the initial join is now set at 90° to the seal — flame anneal and place in a hot oven to anneal fully.

The angle at which the shoulder comes off the seal is most important, the aim is to produce an even compressive seal around the alumina.

I have found this method to produce a robust seal although a straight "end on" join using an intermediary sealing glass can be made to good effect.

The pre-warming for both type of seal has to be equally rigorous. The choice of seal very much depends upon personal preference, what looks right and what has been found to work (Fig. 5.330).

Graded Seals

Graded seals bridge glasses where the expansion rates are vastly dissimilar.

For example, Borosilicate to soda lime glass or Borosilicate to fused Silica.

Because of the type of work I have been involved with, I have used borosilicate to silica most often where an optical cell has needed to be joined directly to a vacuum system. For various reasons which will become apparent later, I have used proprietary brands.

Because of the difference in expansion rates — silica 5×10^{-7} per °C and borosilicate 32×10^{-7} per °C there has to be a series of intermediate glasses with decreasing expansion rates from the borosilicate to match the fused silica.

Each graded glass should ideally have an expansion rate between 1.5×10^{-7} and 1×10^{-7} per °C between each neighbouring glass so that the various intermediary glasses remain robust and intact rather than collapse because of strain caused by the large expansion differentials, even with such a small difference in expansion rate there will always be residual stress within a graded seal.

Fig. 5.330.

There will be four (4) intermediary glasses between borosilicate and silica (Fig. 5.331) and seven (7) between borosilicate and soda-lime glass (Fig. 5.332).

4 ↕	Fused Silica	5×10^{-7} per °C
	Borosilicate	32×10^{-7} per °C
7 ↕	Soda-lime glass	93×10^{-7} per °C

Having a stock of so many intermediate glasses within my workshop was not an option when using only 10–20 graded seals per year, although having an understanding of how to make one was useful when one day I was asked to make a connection between a mass spectrometer and a gas preparation vessel or AGHIS. Having made the vessel which consisted of a 1 L reservoir, a series of solenoid operated valves and a polished flange sample cup — all within a heated chamber — all in borosilicate.

The request was to make the connection using a glass that would not conduct the high level of voltage along the connecting line.

The vessel was borosilicate, the probe into the spectrometer made of Kovar or Nilo K with a glass to metal seal thence via a graded seal to the non-conducting glass and finally to the borosilicate via the seal I had to make.

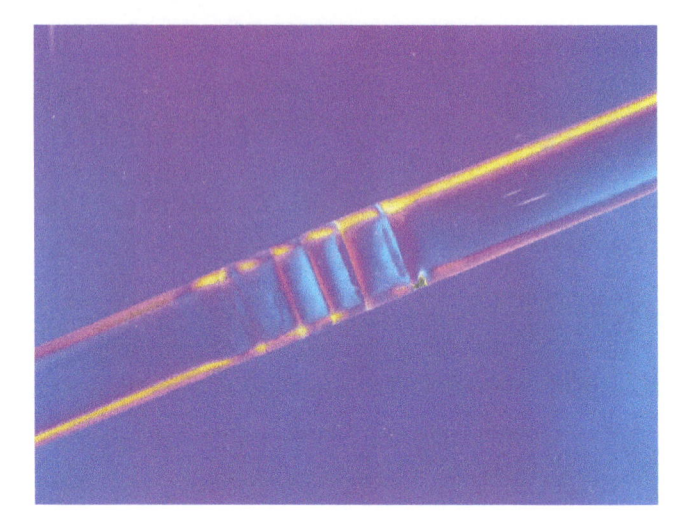

270   *Laboratory Scientific Glassblowing: A Practical Training Method*

Fig. 5.331.

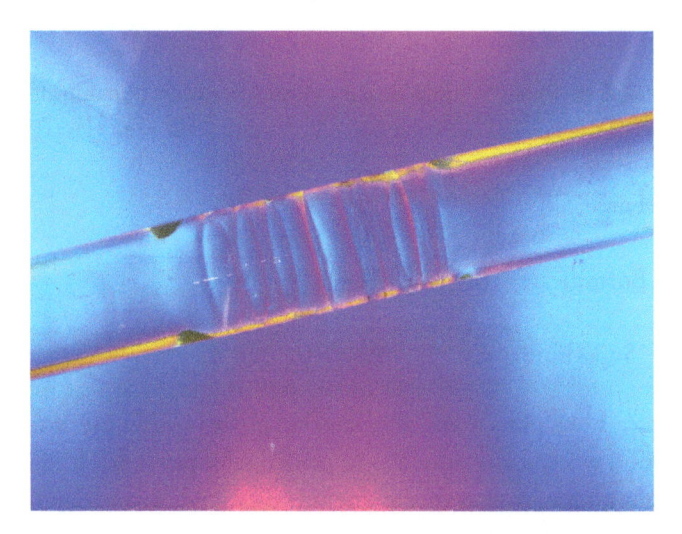

Fig. 5.332.

The graded seal started with the glass of the highest melting point which was set in a lathe via a swivel and blowing tube.

Using a hand torch, the first glass was domed off and blown out to create a small hole. Then using some small diameter rod of the next softer grade of glass, a bead was run onto the very tip and the two glasses "run in" while concentrating the flame on the softer glass which in turn is domed "run in" and blown out and filled with the next softer grade of glass.

Fortunately there were only two intermediary glasses required before the final glass was joined and run in.

A simple job in itself except that the final glass devitrified almost instantly meaning that working with the very tip of a lamina flow flame was the only option to avoid destroying the glass.

I am reluctant to specify each glass because the company who supplied my intermediary glasses was a large electronics company who made their glasses "in house" but are now no longer in business so it would be foolish for me to reference their product codes.

One can only seek specialist glass manufacturers should the need arise to make a non-standard graded seal with regard to the range needed and the price!

5.22 Various Techniques

Davies Double Surface Condenser "The Inner"

Take a piece of 14 mm ø tubing 25 cm long and round one end then blow a dome 15 mm from the rounded end, heat the very tip of the dome and blow while the glass is still in contact with the flame, the hole produced should be reamed to give a hole 10 mm in diameter (Fig. 5.333) while still

Fig. 5.333.

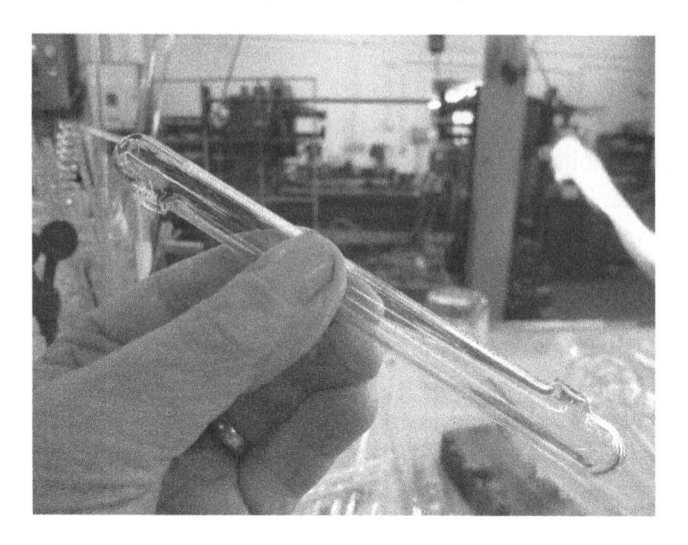

Fig. 5.334.

retaining the raised shoulder. Allow to cool completely, measure 15 cm from the tip of the rounded end, make a mark and then form a rounded end at this point. Make a second dome directly opposite the first 15 mm from the rounded end, once more blow out the tip of the dome while still in contact with the flame and ream out to give a diameter of 10 mm (Fig. 5.334). The inner glass produced will now need to be supported centrally within a length of 24 mm ø tube 35 cm long.

Because of the shouldered holes, it will not be possible to simply wrap corrugated paper around the tube as it would not be held centrally and removal would be difficult. If we continue to use corrugated paper as a support material, folding the paper back on itself will form a channel along which the raised holes may slide, although I would recommend that a glass support be made similar in design to that used to support the spiral when making a condenser in Section 5.15.

A piece of 18 mm ø medium wall tubing which has an internal diameter of 14.5 mm will give good support to the 14 mm inner tube, prepare a 25 cm length then as the 18 ø tube must have a section removed from its length to enable the inner to slide inside. Cut out a section 11cm long from one end. One may at this point taper the 24 ø tubing 8 cm from one end prior to the next stage or after the first side seal has been completed (Fig. 5.335). Using the holder support the inner within the body and warm the tube in a gas/air

Fig. 5.335.

Fig. 5.336.

flame near where the seal is about to be made then adjust the flame to form a fine hot oxygen/gas flame, aim the tip of the flame directly at where the inner hole is positioned, as soon as the outer touches the inner remove from the flame and blow to confirm that a complete circle can be seen to indicate that a seal has been made (Fig. 5.336).

Fig. 5.337.

Fig. 5.338.

Take a fine piece of rod, while heating the very centre of the seal, remove from the flame and touch the centre with the very tip of the rod and pull to form a spike (Fig. 5.337). Run a glass knife across the base of the spike and remove the spike with a tap from the knife. Rotate a fine flame around the edge of the hole — watch as it opens and forms a rounded edge to produce a "side Dewar seal"(Fig. 5.338). If not done before, a slight tapered constriction

Fig. 5.339.

can be made to the right of the inner rounded end (Fig. 5.339). Flame anneal the glass and allow to cool. Remove the support and place the first stage in the left hand, play a fine flame at the centre of the second hole until the seal is made — confirm this by blowing through the first small side hole. Once again pull a spike from the centre of the seal and remove.

"Run in" the edge of the hole then draw a taper to the right of the seal/rounded end. Flame anneal and allow to cool, measure the overall length at 18 cm and mark either end — pull off at the mark and blow out, use a carbon to form a slight flared end in readiness for sealing into the final jacket, allow to cool and repeat this stage at the opposite end (Fig. 5.340). We are now ready to support the two internals within the outer jacket and make a supported internal seal — Sections 5.10 and 5.17.

The outer jacket will be made from 40 mm ø jacket tubing 45 cm long. Start with either a rounded end (Fig. 5.341) or by drawing down onto the seal (Fig. 5.342) rather than sea. The inner section will be supported centrally, the seal made and a suitable size, slightly flanged tube used to make the seal. The side arms of whatever design should be attached and aligned directly opposite the inner section holes (Fig. 5.343). To complete, the packing is removed, the final seal made by blowing down the sidearm and central tube alternatively.

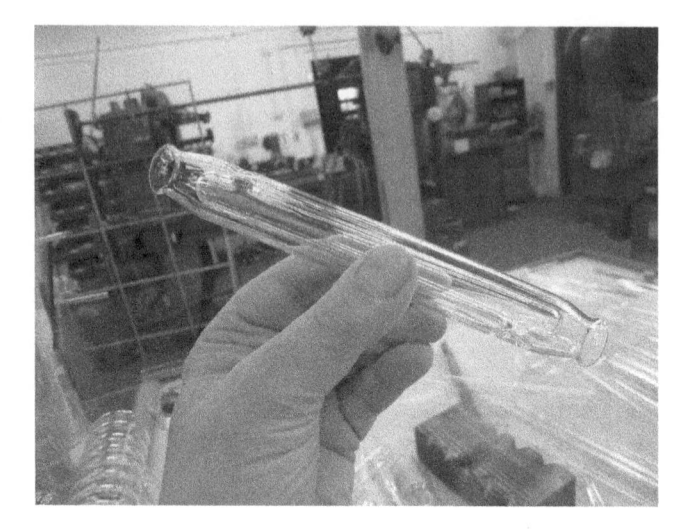

276 *Laboratory Scientific Glassblowing: A Practical Training Method*

Fig. 5.340.

Fig. 5.341.

There has been a great deal of work involved to get to this stage — make numbers of the inner sections before attempting the final seal within the outer jacket (Fig. 5.344). Think the sequence of operations through carefully and have a clear image in your mind of each stage, understand down which tube to blow at any given point and which tube to seal or not to seal.

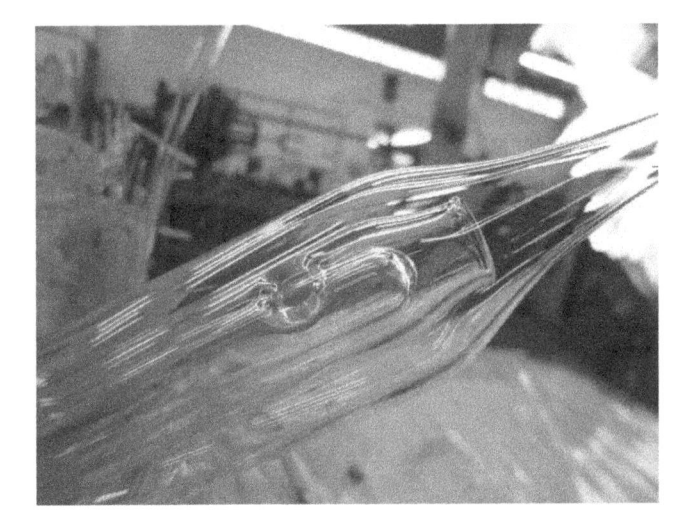

Fig. 5.342.

Fig. 5.343.

Side Arms — Hot, Suck and Tap

When joining a side arm to a heavy walled tube such as the neck of a Büchner flask, the contact point of the side arm has to be as thick if not thicker than the tubing to which it is to be attached. Preparation of the side

Fig. 5.344.

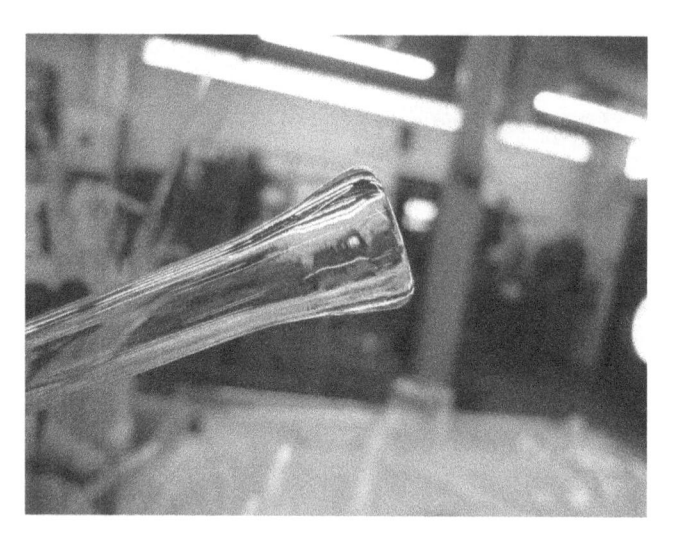

Fig. 5.345.

arm is most important as it must taper out to the seal and be thickest at the point of contact (Fig. 5.345). When the seal is about to be made, both the face of the hole and the side arm must be virtually white hot when the contact is made to ensure that both side arm and hole immediately flow together. There is then an important "pause" before the final blow and pull into shape.

"Suck Seal" is a novel method of making a side arm seal when it seems impossible to do so. The side arm must have a tapered thickened shape.

The Exercise

Take a piece of 30 mm ø tube 40 cm long — clamp horizontally. Prepare a side arm and attach a blowing tube to one end. Using a hand torch, heat the 30 mm ø tube at the midpoint then use a piece of rod and touch the tube and pull some glass away from the place where the side arm is about to be made to thin the wall a little (Fig. 5.346). Hold the side arm directly above the thinned tube and use a hot gas oxygen flame to heat both the side arm opening and the tube until hot enough to glow pale orange in colour. Touch together and immediately "suck" which will pull the tubing inside the side arm and burst! Hold vertically until the glass has solidified while ensuring that the side arm is vertical and at 90° to the tube (Fig. 5.347).

A "brilliant" technique once perfected.

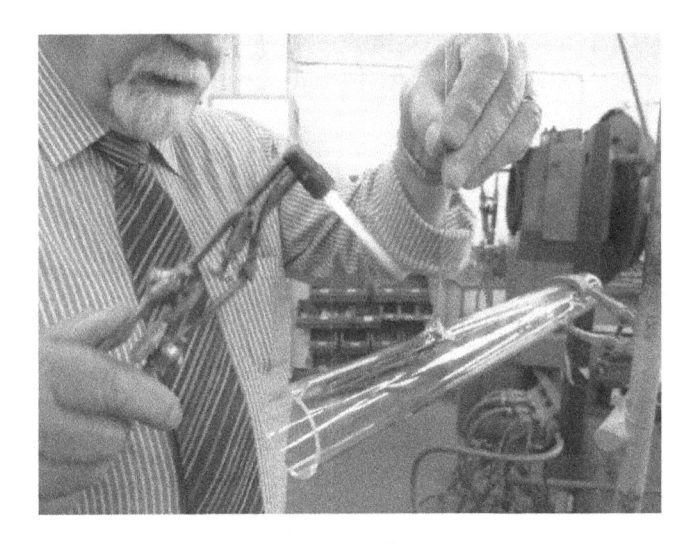

Fig. 5.346.

280 *Laboratory Scientific Glassblowing: A Practical Training Method*

Fig. 5.347.

Fig. 5.348.

Taps — Vacuum Tap Side Arm

The tapered barrel is formed first, the side arms attached, annealed and the internal taper ground before the key is made, pre-ground and then both key and tap are married (ground with finer carborundum grades) together — as each vacuum tap is made individually a "number" is fired onto the corresponding tap and key (Fig. 5.348). A point to note is that keys and taps with dissimilar numbers are "unmatched" and would not function as a vacuum tap as they would most probably leak!

The tap side arms are attached in a way that forms a controlled distortion of the barrel. Allow me to explain.

The Side Arm

The side arm ends are shaped in such a way that having a thickness and bulk will when joined to the barrel push in an indent at the point of contact which on completion will be ground away. The tap is formed with substantial beads at each end — the centre of the barrel measured and marked a rod is attached to the barrel in line with the proposed side arm at right angle to the centre line of the barrel to act as a handle (Fig. 5.349). This is a critical alignment as from this the two side arms depend. A small hole is made in the barrel opposite the "handle" by heating with a fine flame and drawing a spike with a rod which is marked and sliced off with a knife. The hole is opened using a fine carbon point so that the hole tapers towards the centre of the body (Fig. 5.350). A fine hot flame is used to heat both the hole and side arm. The side arm being pushed into the barrel confident that any internal excess will be ground away (Fig. 5.351).

Fig. 5.349.

282 *Laboratory Scientific Glassblowing: A Practical Training Method*

Fig. 5.350.

Fig. 5.351.

Although when produced in volume many aspects of the procedures are automated. To make this type of join at the bench requires practice, practice and more practice to develop the techniques which ensure a clean hole and a smooth strong external join.

Fig. 5.352.

Double Re-entrant — McLeod Gauge

The double capillary join on a McLeod gauge is made to a relatively narrow tube, so there is always a good possibility of the narrow tube distorting. The design of the re-entrants may vary from having the capillary joined directly to the tube which is difficult but not impossible (Fig. 5.352), another design would have the capillary joined at either end with standard tube to make the join (Fig. 5.353). Having first bent the capillary to shape, use this to mark the two join positions. Inline domes should be blown — their position checked and checked again, a small thin bubble should be blown from the top of a dome without it bursting — the second dome in the same way, then each bubble is removed in turn.

The capillary tubing ends should be flared a little (Fig. 5.354) then both the hole and the end of the capillary heated — touched together to make the seal and while still warm the second seal bent into place and aligned — heat the second join and whilst hot return to the first and push the second seal gently together — blow to confirm though this may have to be repeated once more until both seals are made — use a rod to seal large holes! Once together, run in each join in turn while one join gives support to the other until both joins are smooth and well run in. Keep both joins warm by alternating between the two frequently (Fig. 5.355).

284 *Laboratory Scientific Glassblowing: A Practical Training Method*

Fig. 5.353.

Fig. 5.354.

Soxhlet Top Re-Entrant

The large rising vapour tube and final join in a Soxhlet extractor may be completed in the following way:

The top join must be in line with both the lower join and the syphon. Make a judgement as to how long the tube will be required to make the final

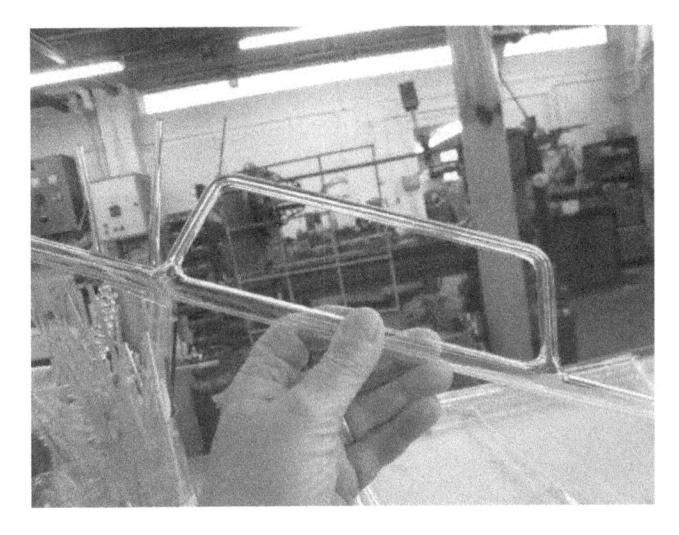

Fig. 5.355.

Fig. 5.356.

bend, estimate where the join will be made, at which point blow a small dome to thin (Fig. 5.356) the proposed point of contact — heat the tubing and pull off the excess glass using tweezers, bend the tube to make contact with the dome (Fig. 5.357) and soxhlet body, aim the flame directly at the

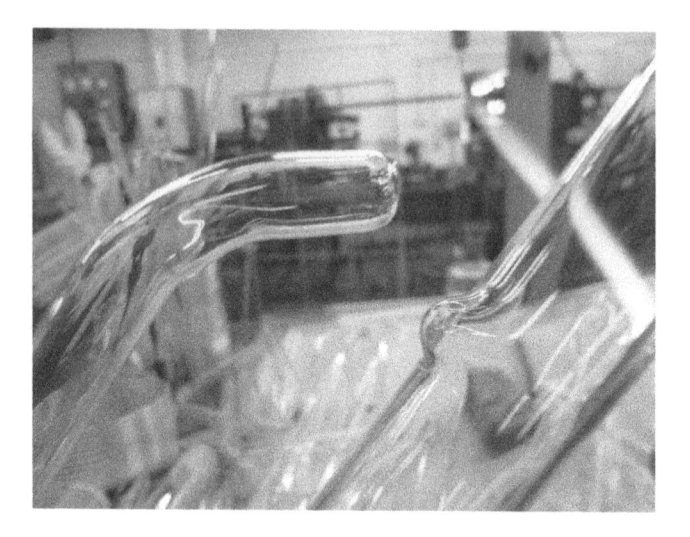

Fig. 5.357.

Fig. 5.358.

point of contact, heat and blow so that the join increases in size which should then show a contact ring and a membrane across the seal (Fig. 5.358).

One will require a sharpened steel point bent at right angle to the stem (Fig. 5.359). Insert the steel point and tap against the centre of the seal.

Fig. 5.359.

Fig. 5.360.

Once broken through, remove any shards of glass and proceed to "run in" the join (Fig. 5.360).

Veigreux Column Indentations

The veigreux column is an air condenser but rather than having smooth sides has a series of indentations to encourage the returning distillate to meet the vapours ascending the column. The indentations are grouped in sets of four at regular distances along the length of the column, some are positioned horizontally others are angled downwards. The tips of each indent almost touch its neighbour.

The Exercise

One will require some fine pointed carbon/graphite points with which to form the indents. Use wet and dry abrasive paper to shape the four-sided "point". Take a piece of 30 mm ø medium walled tubing 30 cm long and use either a wax or fibre pen to mark four equally spaced parallel lines (Fig. 5.361). Then mark lines around the circumference 40 mm apart (Fig. 5.362). One may wish to scribe a small cross at each intersection using the tip of the Glass knife (Fig. 5.363). Although the marks will burn off once the work starts, the remaining lines will give some indication as to the position of the indents and your own estimating skills should be sufficient to keep everything aligned. Working from one end warm the first position then reduce the flame to give a fine hot point; aim at your mark and hold the glass steady while the glass softens. When a circular molten spot is seen, remove the glass from the flame and insert the carbon point directly into the centre of the softened glass (Fig. 5.364). STOP when the carbon almost reaches the centre, withdraw the carbon and heat the tube at a point directly opposite,

Fig. 5.361.

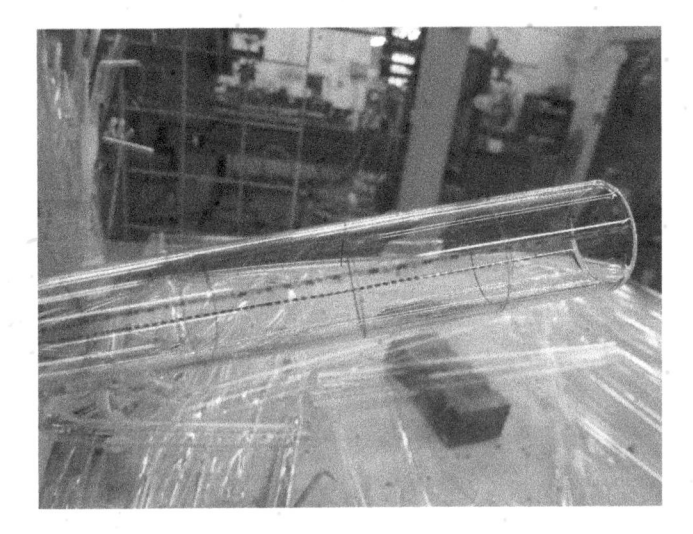

Fig. 5.362.

Fig. 5.363.

once hot insert the carbon in line with the first until almost touching, repeat with the remaining two inserts and warm the area with a feather-like flame (Fig. 5.365). Move to the next position and repeat the operation once more inserting the carbon horizontally. Continue to the midpoint of the tube while warming after each set of indentations and allow to cool completely.

Fig. 5.364.

Fig. 5.365.

Hold the indented section and continue from the midpoint to the end, place in the oven and anneal (Fig. 5.366). Prepare a second tube marked as previously — make the first set of indents horizontally, the second set angled towards the centre, the third set horizontal, the fourth angled, continue to the midpoint before cooling and continue with the same sequence to the end

Fig. 5.366.

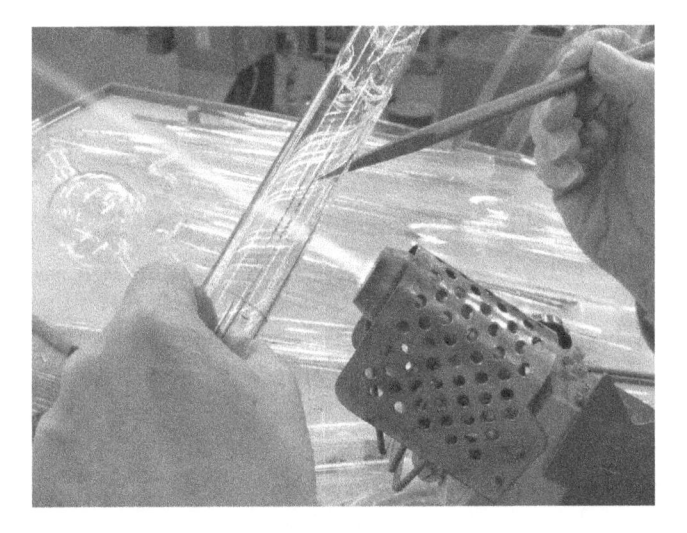

Fig. 5.367.

(Fig. 5.367). As confidence builds, attempt to repeat each exercise without marking off the tube. Inspect the work and note the consistency of the distance between each row of indentations and the parallel alignment. What to look for is consistency.

Platinum Rod — Sealing Holes

Many years ago I was given a piece of platinum 3 mm in diameter the end if which I hammered into a point and have now used for many years to close holes in glass simply by heating the whole area and using the heated point to draw the sides of the hole together without the fear of forming an oxide and contaminating the glass.

When sealing larger holes, molten rod will fill a hole but once sealed the same volume of glass should be removed (reheated and drawn off) and the area reworked to disperse any unevenness in wall thickness.

The Freeing of Joints

Cones and sockets can become difficult to remove from one another due to misuse, chemical attack or lack of lubricant between the two ground surfaces. First attempts to separate joints can begin with light taps on the side of the joint with a short length of wooden dowel, followed by a gentle rocking of the joint.

If at this stage the joints will not separate, we move onto more drastic methods — depending upon certain criteria.

The questions to be asked are as follows:

Are there flammable materials or liquids within or on the glass?

Is the joint/vessel sealed?

Is the joint/vessel under vacuum or pressure?

Does the vessel contain dangerous chemicals?

If the answer is YES to any of the above, further discussion as to how to proceed must take place at the highest level.

If the answer is NO to the above questions, we may proceed to run a drop of methyl salicylate or engineers freeing oil along the rim of the joint and watch as it migrates between the two ground surfaces once it can be seen that the two surfaces have been wetted a quick tap with the wooden dowel and a gentle rocking of the joints should free them (Fig. 5.368).

Often, chemists will bring apparatus with an immovable joint along with the statement "I've warmed it with a hot air gun or I have had it under the hot water for some time and it won't move".

It would seem that all that has happened was that both cone and socket have expanded and contracted at the same rate. If one has to use heat to free a joint the better method is to rotate the joint in the top of a gas/air flame

Fig. 5.368.

Fig. 5.369.

for one or two seconds at the most so that only the socket expands before the cone has time to start, a quick tap with the dowel should prove effective.

Where a cone has broken and remains in a socket, a mild steel bar 12 mm OD 24 cm long with a section removed (Fig. 5.369) has both weight

Fig. 5.370.

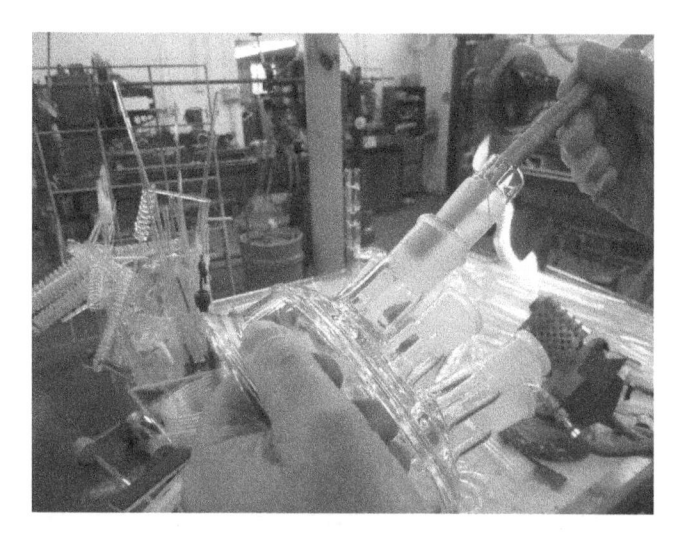

Fig. 5.371.

and impetus when placed inside the cone (Fig. 5.370) and drawn sharply out so that the recessed edge catches the rim of the cone, thus shocking it out of position and freeing it from the joint (Fig. 5.371).

Finally annealing the glassware in the oven may well free the joints from each other — remember to place the glassware in such a way that when being heated in the oven it is encouraged to drop away from the socket rather than being positioned vertically which may allow the cone to drop further into the socket.

Borosilicate Flat Plate

The essential difference between flat borosilicate and tubing or rod is the care and attention needed in pre-warming the material before any attempt can be made to flame work the glass, the pre-warming heat must penetrate through the entire depth of the glass. Once it is hot, borosilicate flat plate is a wonderful material with which to work, it is in many ways similar to working with soda lime glass as one has to use well balanced flames without extreme changes in temperature while using a warming flame between each and every action so as to maintain an even temperature throughout the glass.

Volumetric Ware

Flasks and beakers "contain" whereas the delivery of a specific volume at a specific temperature has a far more rigorous set of controls and regulations to cover the delivery of liquids from burette and pipettes with grades of accuracy as defined by local, national and world standards. One difference between A and B burettes when delivering a particular volume when both are expected to deliver a volume at a specific temperature is that grade A burette has far closer tolerances and also requires a certified "time" for the delivery of the volume.

There are a multitude of websites that automatically calculate volumes of a myriad physical shapes including the volume of a sphere, cylinder or cone at the press of a button. I have often used them myself, the only drawback is that someone else has done the calculation from which you can become disassociated, better to understand the basic mathematics. The need for the correct "mindset" is important especially if you are about to make apparatus of a specific volume or which involves some form of calibration.

Volume of a cylinder

Calculating the volume of a cylinder involves multiplying the area of the base by the height of the cylinder. The base of a cylinder is circular and the formula for the AREA of a circle is

Πr^2 — Π which is a Greek letter and relates to a circle's diameter in relation to its circumference and has an approximate value of 3.14

$$\text{Area} = \Pi r^2 = 3.14 \times \text{Radius} \times \text{Radius}$$
$$\text{Volume} = \Pi r^2 \times h \text{ (height)}$$

A working example: The radius (*r*) is 3 cm and the height 8 cm
$$\text{Volume} = 3.14 \times 3 \times 3 \times 8$$
$$\text{Volume} = 226.08 \text{ cm}^3 \text{ (cubic centimetres)}$$

Volume of a cone

The volume of a cone is one-third the volume of a cylinder which has the same area of base and height:

$$\text{Volume of a cone} = \tfrac{1}{3} \Pi r^2 h$$

A working example: if the radius of the base is 2 cm and the height 7 cm

$$\text{Volume} = \tfrac{1}{3} \times 3.14 \times 2 \times 2 \times 7$$
$$\text{Volume} = 29.31 \text{ cm}^3$$

The volume of a sphere

$$\tfrac{4}{3} \Pi r^3$$

The radius is the distance from the centre of the sphere to the outer edge or half the diameter.

A working example: If the radius is 4 cm

$$\text{Volume} = \tfrac{4}{3} \times 3.14 \times 4 \times 4 \times 4$$
$$\text{Volume} = 267.95 \text{ cm}^3$$

Whenever you become aware of a specific size, length or volume which is relevant to the type or style of glassware, you are currently involved with make a note for future reference, more so make a habit of making notes for future reference.

Nomograph (Fig. 5.372) and sphere volume graphs (Fig. 5.373) are given below.

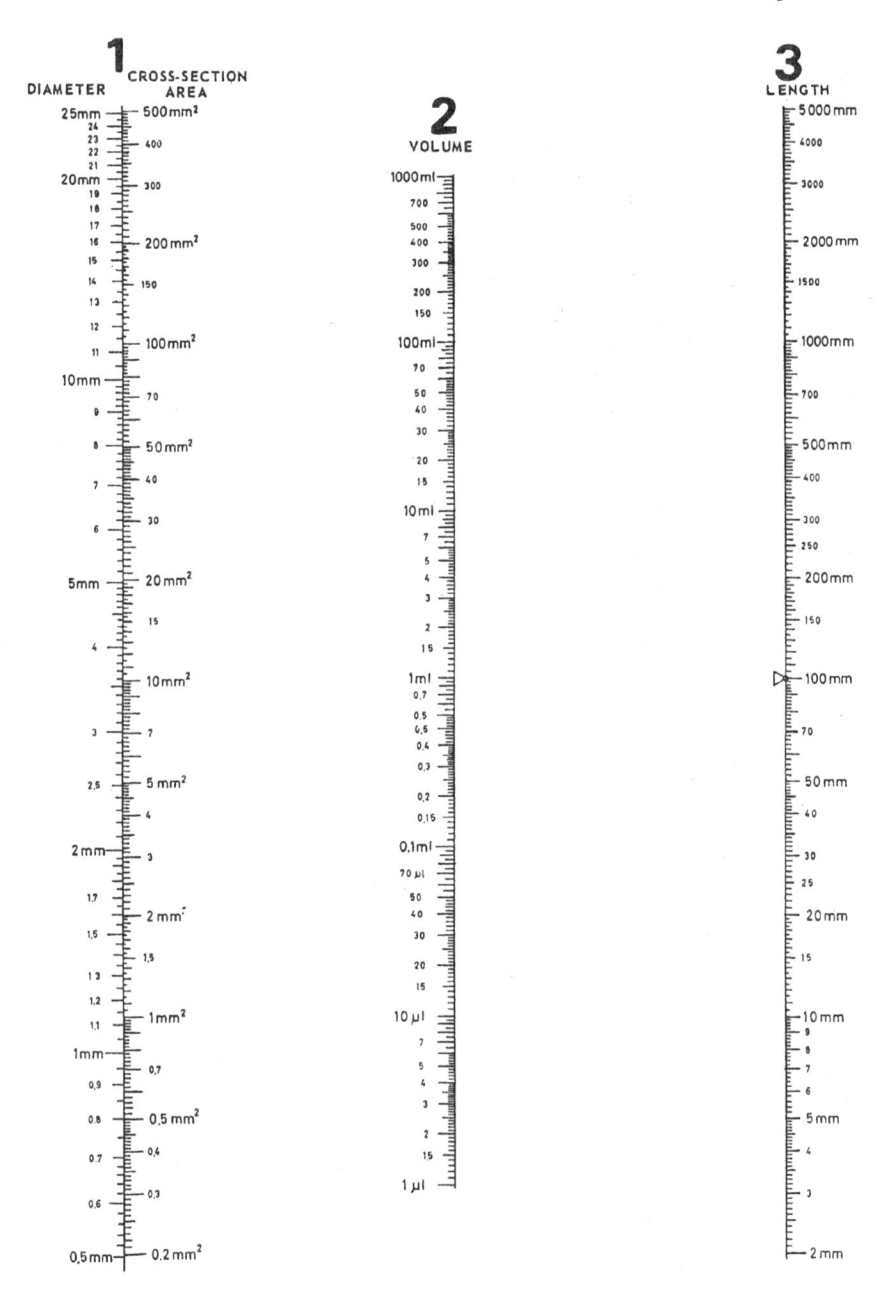

Fig. 5.372. Nomograph.

GRAPH OF RELATION BETWEEN
DIAMETER & VOLUME OF SPHERES

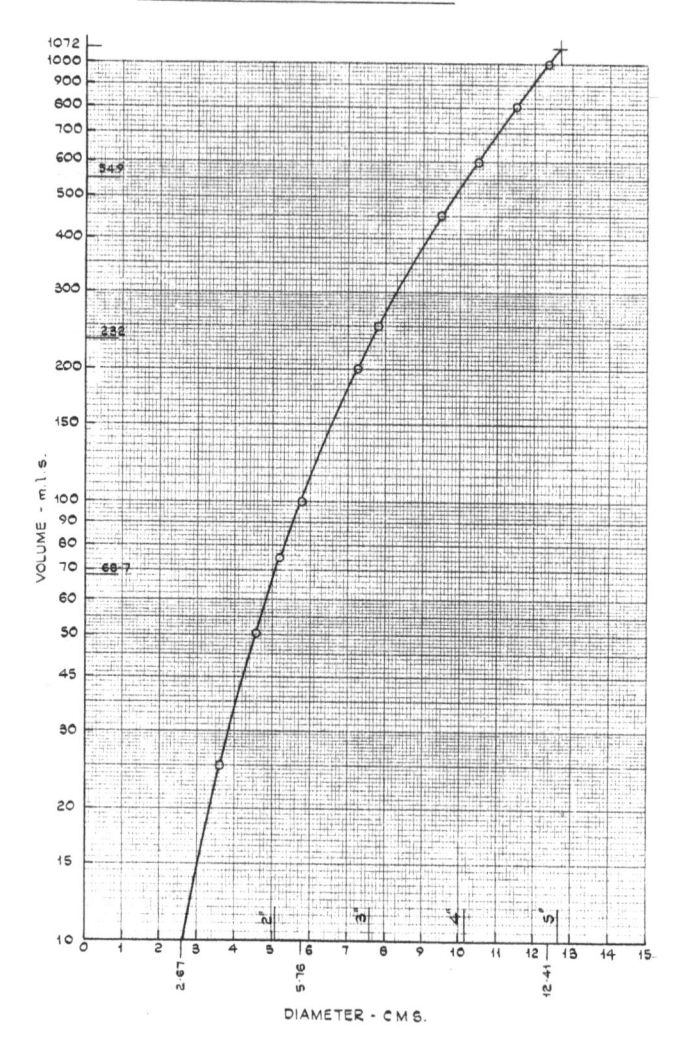

Fig. 5.373. Graph of relation between Diameter and Volume of spheres.

Glass requiring calibration is measured against a Grade A certified volume. The volume is marked from the zero to the total required. The glass surface is then coated with a "resist" then set within a dividing machine where scale divisions are scribed through the resist which is then acid etched using 60% hydrofluoric acid for a set period of time. The acid is neutralised, the resist removed and the resulting etched grooves filled with glass enamel, when the glassware is next annealed, the enamel fires directly into the lines.

I have no intention of describing in great detail the etching process as I feel it would be totally irresponsible for me to do so.

Because hydrofluoric acid used in the etching process is extremely dangerous both in its liquid form and its vapours are toxic causing immediate and serious burns. The acid reacts readily with all forms of calcium including bone!

The safety equipment needed has to be of the highest standard and the etching operation conducted within a fume cupboard. Training in the use of dangerous chemicals has to be of the highest order. Lone working is forbidden and disposal of residual acid highly controlled and monitored.

Units — Their Metric History

The units used in modern calibrations have their historical base in 1793. It was then that the litre was first defined as a liquid measure, a convenient volume for commercial use. Towards the end of the nineteenth century precise measurements were required in chemistry, physics and engineering and in 1889 the "standard kilogram" was constructed, which was intended to be the same mass as 1 litre of pure water at the temperature of its maximum density, 4°C. The litre was then officially defined as the volume of 1 kilogram of pure water at 4°C. It was not until 1907 that a slight error was corrected. The 1889 "standard kilogram" was found to have a mass of 1000.28 cc of pure water at 4°C and it followed that the litre was 1000.028 cc.

It was decided to leave the kilogram as the "standard" but to divide the litre into 1000 equal parts and to call this division by a new name: the millilitre (mL). One millilitre then equalled 1.000028 cc and 1 cc equalled 0.999972 mL. So from 1907, millilitres were used as the standard unit of liquid and volume measurement. In 1964, the General Conference of Weights and Measurements redefined the litre as equal to 1000 centimetres cubed (cm^3 or cc). This means that the litre by its new definition is directly related to the metre as a measurement of volume and no longer to the kilogram. However, vessels calibrated in millilitres (mL) are likely to be found in use for some time to come, except for very accurate analytical work it may be assumed that 1 millilitre (mL) equals 1 cubic centimetre cm^3 or cc.

An Exercise

In Preparation for this manual, I have referred to my notes from 1966 when I was undergoing my initial training.

From the diagram (Fig. 5.374) one can see that the exercise consists of a jacketed body with side arm and cross piece with a re-entrant. My notes say that the exercise was made after having four months training.

I was informed that all dimensions were critical and that the whole exercise would be made in soda lime glass. It may well have been an exercise in balancing temperatures with this type of glass but I know now that it was an

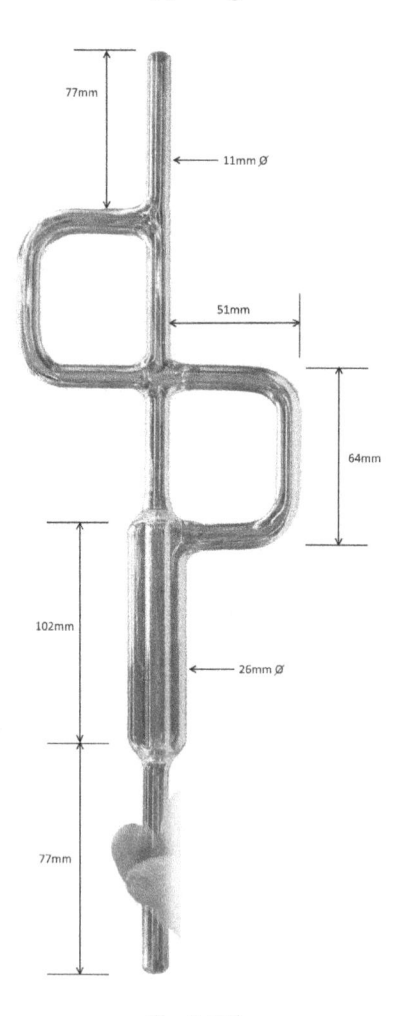

Fig. 5.374.

exercise to see if I had the persistence to continue despite the inevitable failed attempts that would follow.

Complete the exercise in borosilicate and then if the opportunity arises in soda lime glass, if you are successful in "Soda", I hereby metaphorically pat you on the back and say "well done".

The Naming of Parts

The body is made from 26 mm ø tubing into which the inner 11 mm ø tube will be supported with corrugated paper after which it will need to be removed. We have a body length of 102 mm and a tail of 77 mm, a nominal overall length of 180 mm or 18 cm.

Take a piece of 26 mm ø tubing 36 cm long, mark the centre, heat and draw a spear point at the mark, burn through the spear — heat the shoulder after taking off the spear and make a domed end.

Support a 10 cm length of the 11 mm ø tube centrally within the body.

Cut two further lengths of the 11 mm ø tube 10 cm long for the bottom stem which will be cut back to 77 mm when the second seal has been completed.

Two 11mm ø re-entrant tubes will need to 260 mm long, that is, 55 mm plus 55 mm plus 64 mm and a further 85 mm with which to make the two bends and also have sufficient excess as to pass the cross piece, align and be cut back to the correct length.

The long stem will need to be 240 mm long that is: 64 mm plus 64 mm plus 77 mm plus a further 45 mm excess which will be cut back to 77 mm on completion of the exercise.

Cut spare lengths of tubing for each section, sharpen you glass knife before starting, have sufficient corks and your callipers to hand.

Draw some rod down to 2 mm ø just in case you need to fill in one or two holes at the joints.

Take the prepared body and carefully warm the rounded end, once hot reduce the size of the flame and heat the very end, watch as the glass melts back to make contact with the 11 mm ø inner, once contact has been made withdraw from the flame and blow to confirm the seal (Fig. 5.375). Re-heat the seal and blow out the centre (Fig. 5.376) and join the 240 mm 11 mm ø tube, run in the join and centralise the longer tube as it cools

Fig. 5.375.

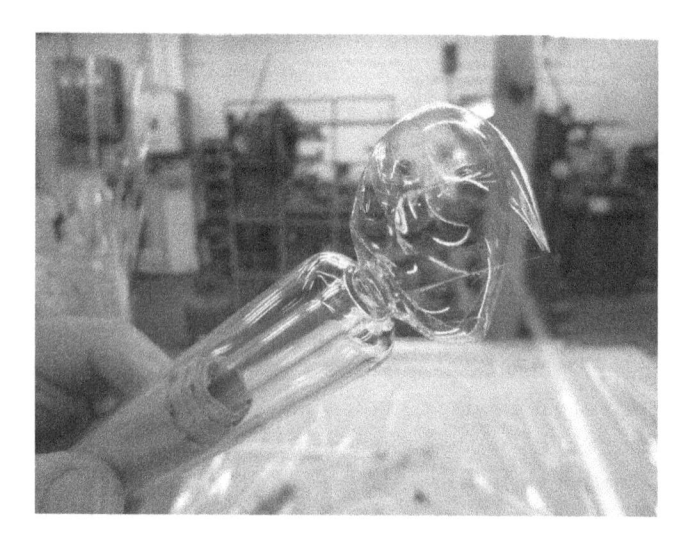

Fig. 5.376.

(Fig. 5.377). Blow a hole at the shoulder (Fig. 5.378) then take a length of the 260 mm by 11 mm ø tube and make a side arm connection, make allowance for the body diameter and make a 90 degree bend in line with the body and the long tube so that the final position will be 55mm from and parallel to the longer tube (Fig. 5.379).

Fig. 5.377.

Fig. 5.378.

Mark the longer length at the cross piece position (Fig. 5.380) allow for the width of the tubing and blow a dome but do not blow it out (Fig. 5.381) Make the second bend aligned to the dome, the excess tubing passing close to one side of the dome, carefully mark and use the "hot point" method to cut the tube to length, heat and pull a small point from the tip of the dome,

Fig. 5.379.

Fig. 5.380.

cut off and open out to 11 mm diameter, warm the second bend and as the glass softens align the tube using tweezers.

One may or may not be successful in making a close connection which can be heated, sealed and run in while occasionally warming the bend to pull the tube onto the seal. Run in and realign the bend and re-work to

Fig. 5.381.

Fig. 5.382.

shape all of which can be quite satisfying as everything is self-supporting (Fig. 5.382).

If one needs to seal small gaps with rod, remember to remove excess rod after the hole has been sealed. If you decide to blow the hole rather than pull a spike, take great care when marking and snap cutting the tube to length (Fig. 5.383).

Fig. 5.383.

After having completed the first part of the cross piece, it would be necessary to continue with making the cross and then the final seal if using soft soda lime glass while constantly balancing the seal temperature, flame annealing and allowing to cool before withdrawing the packing and making the second seal and bottom tube to complete. This would mean that the left hand would be holding the long tube 220 mm to the left of the seal.

When using borosilicate and having made half of the cross piece we can withdraw the packing and make the bottom seal (Fig. 5.384) then attach the bottom tube (Fig. 5.385) which will be so much closer to the left hand and far easier to handle, confident in the knowledge that the cross piece can be "warmed" (Fig. 5.386) and completed (Fig. 5.387) without the fear of it cracking, which would definitely be the case if using soft soda glass.

The completion of the cross and re-entrant will again be self-supporting once the initial joins have been made which enables the joins to be reworked and the bends aligned and smoothed out.

Care must be taken not to overheat the longer tube which might soften and move out of line, good flame control and the occasional flame annealing

Hand Techniques 307

Fig. 5.384.

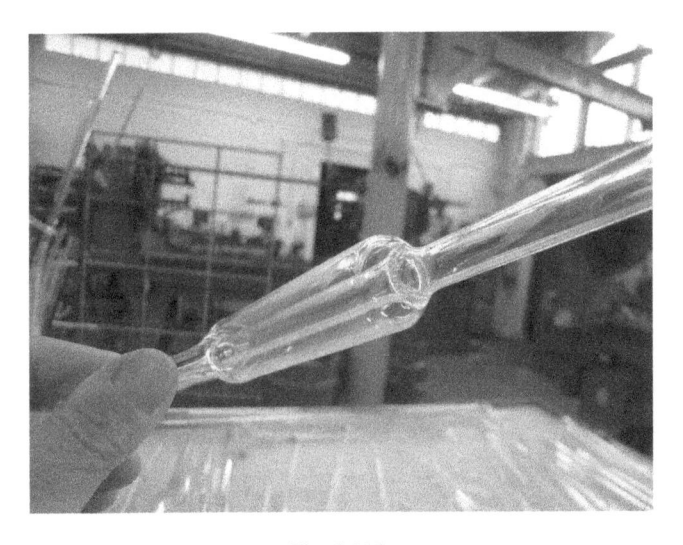

Fig. 5.385.

Fig. 5.386.

Fig. 5.387.

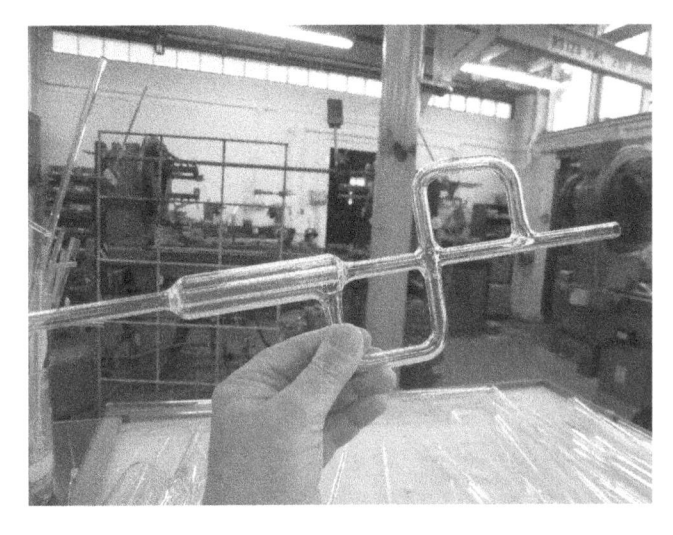

Fig. 5.388.

to disperse any heat-induced strain will be beneficial. Finally, cut both end tubes to length and glaze the ends (Fig. 5.388).

Once you have completed the exercise, anneal the glassware, wrap it up and take it home, place the glass in a box and leave it in the attic for twenty years after which time take it down and "read the glass", look closely at the wall thicknesses and for the signs of where you may have had a problem.

Make another one and compare how your hand skills have improved with practice, confidence and muscle memory.

6

Hand Torch Techniques

There was a time not so long ago when every laboratory had bunsen burners, naked flames and even glassblowing benches. PhD students were expected to construct their own vacuum lines as part of their thesis (sometimes with the assistance of the resident Scientific Glassblower) which fostered good relationships that lasted for years.

Gone are the chromic acid baths at the end of each laboratory bench for cleaning glassware; also, the delightful sweet smell of marzipan as you walked the corridors of the Organic chemistry department which heralded the removal of "Benzene". From that day forth all the other solvents in daily use such as carbon-tetra-chloride were confirmed as dangerous to health and were to be replaced by safer water-based decontaminating solutions. The removal of naked flames from the laboratory led to a reduction of fully glassed vacuum systems using mercury diffusion pumps and mercury manometers.

I have been asked in recent years to work with fully glassed systems in very specialised situations. As this type of work has reduced considerably, many systems are now demountable, flexible, while using grease free taps and electronic rather than mercury vacuum measurement and to produce the high vacuum turbo molecular pumps, all of which means that there is less opportunity or need to repair vacuum lines *in situ*.

As I have aged and become more curmudgeonly I make and repair vacuum lines and the like in the workshop where vac lines can be turned upside down to access a join rather than struggle from underneath, all within an

Fig. 6.1.

environment without restrictions on naked flames and with access to a good size annealing oven. All of which brings me to the type of burner required to assemble a vacuum line — which is a hand torch.

There are various designs and sizes, the choice depending on availability, price and preference. Some have the jet coming off the handle at 30° which is fine when used for lathe work but when constructing a vacuum line or the like, I have a preference for the flame to come away from the burner at 90° (Fig. 6.1) and am quite prepared to bend the neck of a burner and make it do so!

I feel that the glass should be at eye level when making a right angle join to a manifold so as to see the join being made as the glass flows to create a pool of glass at the join, all the easier if the burner is held vertically and twisted from the wrist back and forth so that the flame reaches to the rear of the seal.

Using an angled burner head to do the same task would cause the elbow to rise causing the flame to move in a far greater arc.

Having got that off my chest, let us proceed by taking a length of 30 mm Ø tubing 50 cm long, seal one end with a cork, the other a cork and blowing tube via a swivel or bend so that the possibility of the weight of the blowing tube kinking and sealing off the tube is reduced (Fig. 6.2).

Prepare ten 15 cm lengths of 12 mm Ø tubing one end of which is to be sealed with a cork. Mark the centre of the 30 mm Ø tube and clamp the tube

Fig. 6.2.

horizontally at both ends just sufficiently to hold the weight but with the option of being able to rotate the tube with some resistance from the retort clamps.

Exercise 1

The glass is not going to move only the flame. Warm the large tube at the mark using a soft flame then reduce it in size to give an oxygen/gas flame 10 mm wide, heat the mark with the tip of the flame until it is seen to begin softening at which point remove the flame and blow to create a dome, then as an exercise reheat the dome and blow a thin bubble which can be sliced off to create a hole, smile as you watch the thin bubble float down sometime sticking to the area around the hole!

The better way I would suggest would be to use the hot blow out method by resting the flame on the dome as you blow hard to make the hole, either method created hole can be resized with a carbon reamer, as we have made the hole using the first method we will use it for the sake of this exercise.

Take the 12 mm Ø tube in the left hand; hold it in line with the hole slightly inclined to the rear as you play the flame across the hole which

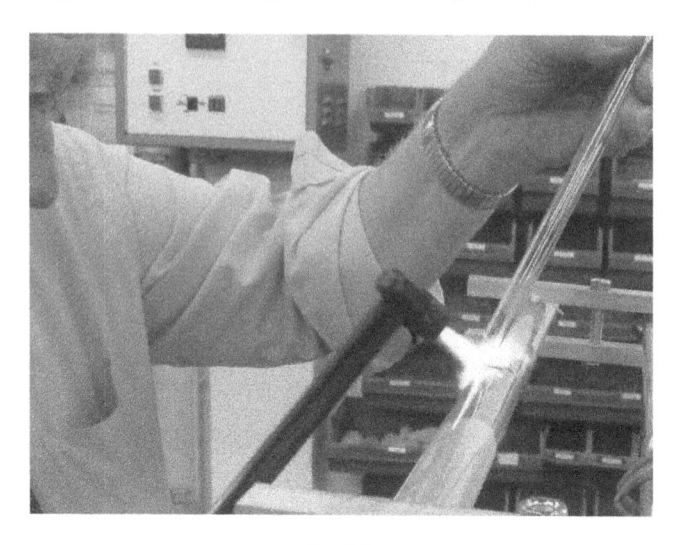

Fig. 6.3.

should also heat the open end of the tube (Fig. 6.3). When seen to be molten, remove the flame and touch the two edges together, pull and blow to make a seal, if all is well reduce the size of the flame and immediately rotate the flame using a twisting action of the wrist while making sure that the flame stays in contact with the glass at the point of the initial join, making sure that the rotation starts and finishes on the far side of the join — hold the 12 mm Ø tube lightly but vertically, the weight of the tube will feed into the seal; once a complete ring of molten glass can be seen at the seal — withdraw the flame — hold the glass vertically and in line, breath into the glass, watch as the shape of the join forms as the outer skin of the glass starts to cool.

Blow hard, pull slightly to realign if necessary, be aware of the wall thickness — allowing the 12 mm Ø to become part of the join will result in a substantial join without the fear of losing the wall thickness (Fig. 6.4).

Exercise 2

Warm off the join with a soft flame, move seven centimetres to the left of centre, make a hole and repeat the exercise, ensuring that the two verticals are in line. Warm the 30 mm Ø tube once more in a soft flame until flecks of the sodium flame are seen. Move 7 cm to the right of the centre, prepare

Fig. 6.4.

a hole and make a vertical join. Remove the cork from the centre tube. Take a piece of 6 mm Ø rod, heat the end of the 12 Ø tube and the 6 Ø rod, touch the end of the tube and draw it away to form a vertical spike or small spear point, burn this off and heat the very end of the sealed tube — blow to make a rounded end, repeat with the right hand vertical.

Exercise 3

Measure the distance between the uprights and cut a piece of 6 mm rod to that length (Fig. 6.5), use a pair of tweezers to hold the rod, heat the end of the rod, touch the hot glass near the rounded end of one of the uprights so that it lies horizontally between the rod and upright, use some more rod to fill the gap by heating the end, pull a small spear, heat the end of spike and melt it directly onto the rod and tube melt the two together — practice making a smooth join. If the rod tip is within 1 mm, use a small flame to heat the tube directly in front of rod; withdraw the flame — blow gently to form a bulge which should touch the tip of the rod — run in the join with a small flame remembering to heat both sides of the area — back and front. Do not overblow. We have braced the two uprights so they will now give support to each other (Fig. 6.6). Look closely at the original joins, re-heat and remelt

Fig. 6.5.

Fig. 6.6.

one of the joins, you will find it much easier than before as the support given by the brace will allow the running in process to be made without worrying about misalignment or losing control (Fig. 6.7). Work on the join either all the way round or section by section. The glass can be inverted, try working in this fashion while concentrating on the wall thickness (Fig. 6.8). Repeat the

Fig. 6.7.

Fig. 6.8.

exercise until confident and have managed to stop catching your hand with the flame or touching the hot glass.

Exercise 4

Set up a piece of 30 mm Ø × 50 cm long tube in the clamps as the previous exercises. Take a piece of 12 mm Ø tube 40 cm long, mark the centre and make a U bend at this point with the parallel tubes between 8 cm and 10 cm apart, allow to cool. One leg will no doubt be longer than the other — cut the longer leg so that each leg is the same length (Fig. 6.9). Make two or three spare U bends. Mark the centres of the parallel tubes on the 30 mm Ø tube with a fibre pen then confirm the mark by scratching a small cross with the tip of your tungsten blade. Warm the tube, reduce the size of the flame and form a dome at each marked position (Fig. 6.10) making sure that they align with the U bend tubes. Heat the tip of each dome in turn, do not blow them out until both have a thin bubble (Fig. 6.11), then and only then slice them off and ream the holes with a carbon so that both holes are in line with the U bend and of the same diameter (Fig. 6.12).

Hold the U bend over the holes, rest the far side on each hole (Fig. 6.13) while you quickly move the flame from each hole every 2 seconds until both

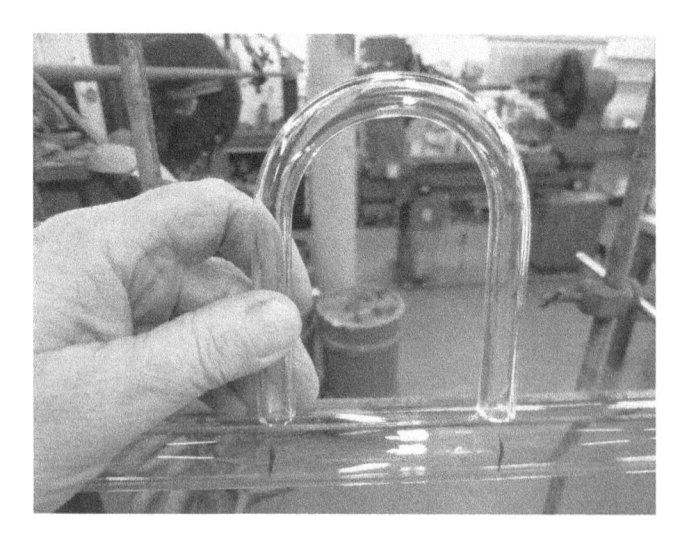

Fig. 6.9.

Fig. 6.10.

Fig. 6.11.

are hot, pull the U bend forward so that contact is made then into the vertical, using a small sharp flame heat each seal alternating between the two while blowing to confirm the seal is made (Fig. 6.14).

One may be successful in part where a hole has remained on one or both seals, draw a rod to a point and fill in the hole with the molten tip and burn

320 *Laboratory Scientific Glassblowing: A Practical Training Method*

Fig. 6.12.

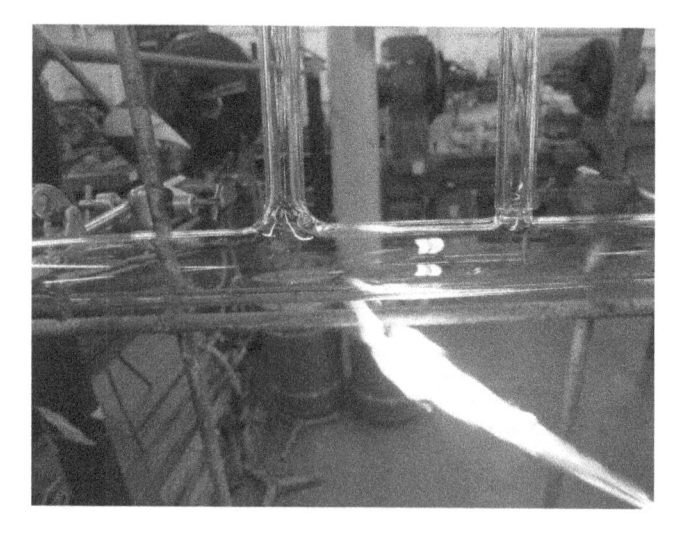

Fig. 6.13.

off the excess, confirm that the seal is complete while occasionally keeping both seals warm — heat the place where the rod was used to fill in the hole, use the rod once more to remove the same volume of glass that was added. Run in each join in turn, firm in the knowledge that one seal will support

Fig. 6.14.

Fig. 6.15.

the other (Fig. 6.15). BUT remember to keep the area warm so as to balance the stresses which inevitably are building up within the glass.

Repeat the exercise until it stops cracking as it is important that this two-point seal is mastered.

Fig. 6.16.

Exercise 5

Make up two wooden clamps as per diagram. Cut two lengths of 30 mm Ø tube 50 cm long — mark the centre of one and blow a hole, anneal the area in a soft flame, clamp the two tubes parallel to each other with the tube with the hole to the rear and the hole on the top, seal all ends with corks having the blowing tube set in the front tube.

The clamp has cut outs so that the tubes will be set apart at a known distance. Using 12 mm Ø tubing make some U bends with their centres to the known distance (Fig. 6.16). Make a hole in the front tube directly in line with that on the rear – confirm that the U bend arms will align. Hold the U bend in the left hand, incline slightly to the left and heat both seal areas, once hot, touch together to make the seal — blow to confirm, close any holes if present (Fig. 6.17). Move from each join in turn to "run in" while occasionally warming both seal areas to balance the strain. Once the seal has been run in confirm the alignment of the U bend (Fig. 6.18).

Repeat the exercise along the length of the tubes — you have in essence constructed a dual manifold Schlenk line — anneal it before the glass falls apart but take a quick look at the amount of strain present using polarised light.

At this point, it would be beneficial for you to return to the first exercise simply to compare your work with the first attempts.

Hand Torch Techniques **323**

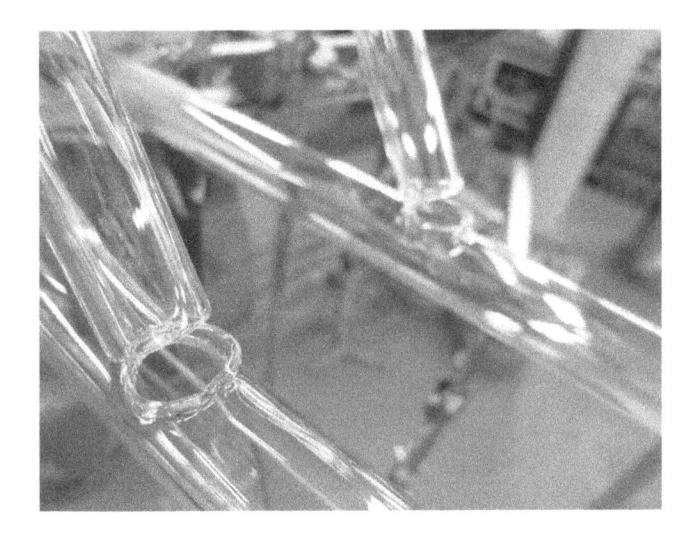

Fig. 6.17.

Fig. 6.18.

Fig. 6.19.

Exercise 6

Prepare a 30 mm Ø tube 50 cm long; attach the blowing tube and cork. Take a piece of 12 mm Ø tube and either grind or cut a 45° angle on one end, after washing and drying the tube seal the opposite end with a cork (Fig. 6.19). Mark the centre and create a hole to which the join will be made at a 45° angle. The technique is the same as for the vertical join although there is a tendency for the inner angle to thicken.

The more important part of the exercise is the left hand grip; make the join while having the tubing angled to the left then a second 7cm to the left of centre and attach so that the two tubes will be parallel and angled once more to the left without bracing the two tubes.

Move 7 cm to the right of centre and attach a third tube angled this time to the right, then a fourth 7 cm further to the right — all should be inline and you should have prepared at least six lengths of the 12 mm Ø tube (Fig. 6.20).

Before moving on to the next exercise, prepare a piece of 30 mm Ø tube and clamp it at 45° (Fig. 6.21), prepare a number of 12 mm Ø tubes, mark the centre and blow a hole, attach one of the 12 Ø tubes (Fig. 6.22) which will now be held vertically, make a further two joins above and two below the centre.

Fig. 6.20.

Fig. 6.21.

I have not described in detail Exercise 6, only its requirements. You will by now have realised that it is an exercise for the control of the left hand which has to react when holding and supporting the tube with an independence and finesse, while the right hand contorts and controls the flame to make the angled join, the inside of the angle being the most difficult part of

Fig. 6.22.

the operation requiring a fine flame while the opposite side requires a larger. It is the change of flame from one to another while maintaining an even melt that taxes ones dexterity.

Exercise 7

Utilise the glassware from a previous exercise after it has been annealed in the oven. Clamp horizontally and seal all ends with corks and a blowing tube.

Take a piece of 12 mm Ø tube 25–30 cm long, make a T piece join and run the seal in, warm the main 30 mm Ø tube to reduce the stress.

Hold the upright in the left hand, soften the flame and stroke the tube over a length of 5–10 cm, as soon as the glass is felt to be softening remove the flame and gently put pressure on the glass to move to the left; as you feel the glass solidifying reheat the tube once more with the soft flame while ensuring that the glass on the outside of the bend receives a little more heat — remove the flame and repeat the encouragement of the glass to bend while blowing so that the tubing does not collapse. Repeat once more only heating the glass until flecks of the sodium flame appear.

With practice, one can make a 90° bend from the vertical, the tubing may well have constricted; let the glass cool, then heat only the inside of the bend without allowing the flame to remain stationary for any time. Blow to regain the diameter. Repeat the exercise but make the bend in the opposite direction. Once again the left has to work independently in a controlled manner.

Exercise 8

The hand torch is an integral piece of equipment for lathe work especially when attaching inlets/outlets on jacketed vessels, prewarming glass and interwarming (keeping the glass up to temperature mid operation).

Take a piece 50 mm Ø tubing, seal and attach a blowing tube and swivel. Mark the centre and place in a lathe; start the lathe and heat the centre mark, once hot use a carbon to constrict the diameter by 2/3 (Fig. 6.23). Reduce the size of the flame, heat the constriction and pull the right chuck away as you burn through the constriction (Fig. 6.24). Remove the glass from the right chuck. Remove any excess glass from the end of the tube, reheat to form a domed end (Fig. 6.25), soften the flame and warm off the area of the dome and tube.

Take a piece of 20 or 25 mm Ø tubing 20 cm long and seal one end. Start the lathe once more and warm the dome and 10 cm of the adjacent tube until flecks of the sodium flame appear. Stop the lathe and switch it off, using the hand torch blow a thin bubble of glass (Fig. 6.26) 10 cm from the end of

Fig. 6.23.

328 *Laboratory Scientific Glassblowing: A Practical Training Method*

Fig. 6.24.

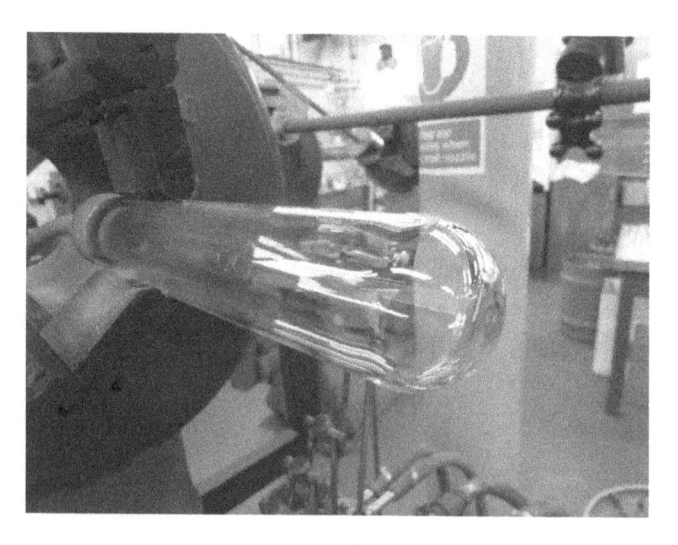

Fig. 6.25.

the dome. Grip the chuck and rotate the lathe by hand so that the bubble is below the tube and slice it off thus avoiding the thin glass touching and sticking to the area to be worked (Fig. 6.27).

Return the hole to the top; it may be best now to use a heat proof glove on the left hand. Using a gas/oxygen flame as wide as the tube and hole, melt

Fig. 6.26.

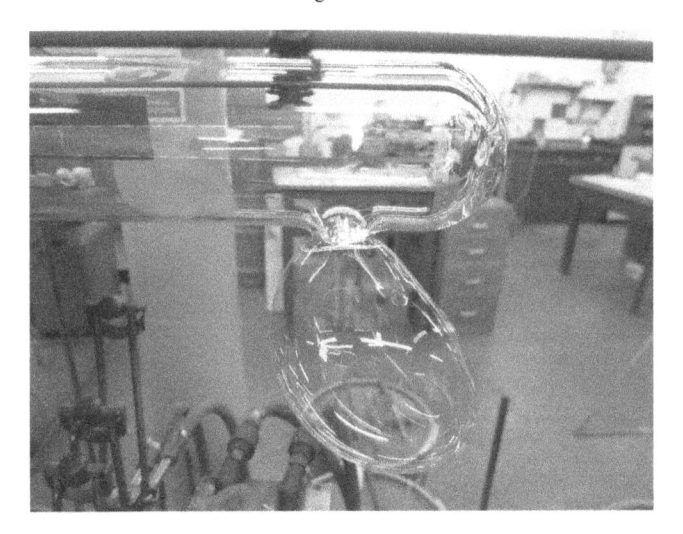

Fig. 6.27.

both edges then touch together to make the seal, pull slightly and blow to confirm the seal, align the tube at 90° in both directions (Fig. 6.28).

Run the join in by either rotating the flame at the base of the tube while concentrating on getting to the far side of the join or by section heating about a quarter of the seal at a time while allowing the rest of the cooling seal to give support.

330 *Laboratory Scientific Glassblowing: A Practical Training Method*

Fig. 6.28.

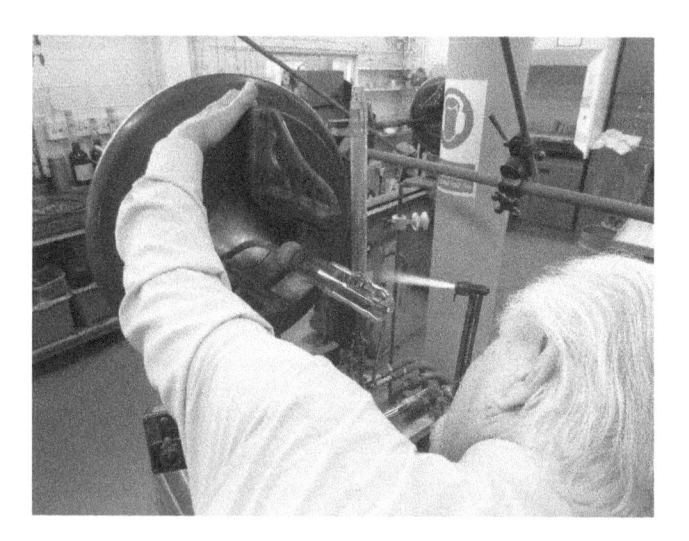

Fig. 6.29.

Once satisfied with the join, confirm the alignment (Fig. 6.29) while making sure that the right hand chuck has been moved away so as not to interfere with the back of your head.

Once again it is the lightest touch of the left hand that is needed when the join is in need of stretching a little, slide your fingers along the glass in a stroking action as the glass cools and solidifies.

To finish this exercise take a B19 socket, scribe a mark with your glass knife 30 mm from the end of the ground surface and using a "hot point" crack off the stem.

Use a B19 cone as a holder with corrugated carbon between the two surfaces, place a cork in the end of the B19 cone.

Whilst the 50 mm Ø tube is still relatively warm — re-heat the whole area of the previous seal, grip the chuck and rotate so that the previous join is below with the side arm pointing vertically down, heat and blow a hole directly opposite the previous side arm (Fig. 6.30). Then seal the B19 socket to the tube in just the same manner as previously, then confirm the join is vertical, remove the B19 cone and the corrugated carbon packing and warm the whole area to reduce the thermal stress (Fig. 6.31).

Give yourself a pat on the back after first putting the hand torch down.

One advantage of making this type of join on the lathe is that the glass can be rotated so that one can see just how well the far side of the join has been run in, which makes it far easier to be reworked if necessary and for the glass to flow back into the sidearm (Fig. 6.32).

Hand torches come in a variety of sizes, the larger types are ideal for prewarming due to their size and shape of flame they can produce. A suitable stand enables prewarming to take place without the need to hold the weight of the burner for long periods as they can be easily removed and

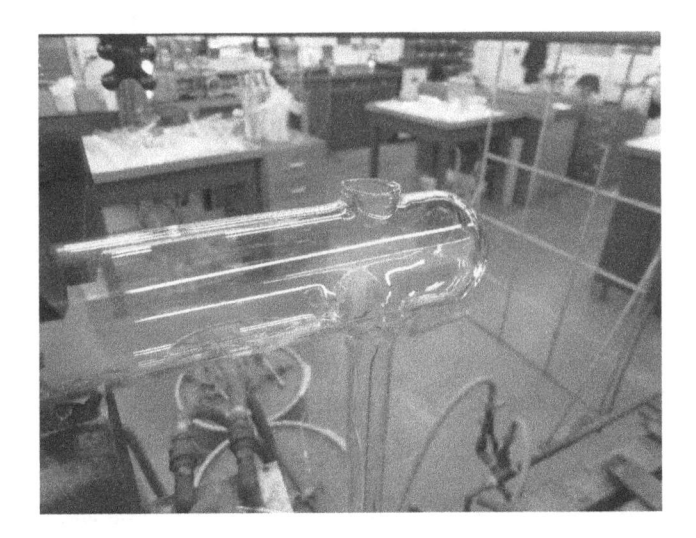

Fig. 6.30.

Fig. 6.31.

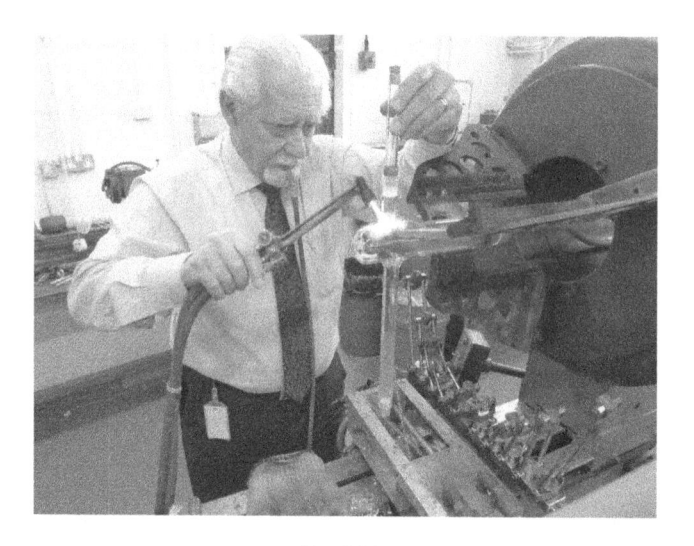

Fig. 6.32.

aligned "end on" to the work (Fig. 6.33). They can give much needed time to remake that missing joint while it keeps the glass warm in a large feather-like flame (Fig. 6.34).

When attaching braced joints to a flange lid, the flame will recover the temperature of the flange or buttress between each joint made.

Fig. 6.33.

Fig. 6.34.

When working on flat borosilicate plate, it is so important to warm the glass thoroughly, the use of a large burner on its own stand is ideal in that you are not pushed to start the work before the glass is ready.

I have found Borosilicate plate to be a wonderful material to work once it has been prewarmed correctly and the exact opposite if one tries to rush making a join to it too early.

7

Vacuum Manifold — Vacuum Measurement — Schlenk Lines

What is vacuum? A dictionary definition may well say a space completely devoid of matter, a space or vessel from which the air has been exhausted to the furthest possible extent by a pump or such means below atmospheric pressure.

As this planet travels through space rotating as it goes with both night and day temperatures, all of these have an effect on atmospheric pressure and also vacuum measurement.

The concept of vacuum in glass can be daunting. Think of a vacuum as a negative atmosphere which can mean that on an evacuated glass vessel a pressure of no more than 14.7 pounds per square inch is pressing on its outer surface. Consult the safe working pressure chart for a given diameter, although designed for internal pressure the safe working pressure should reassure one that they are far and above the 14.7 pounds per square inch, suggesting that a glass vessel is strong enough to withstand one atmosphere (Fig. 7.1).

"But" and it is a big "but", the shape of the glass is of paramount importance requiring a consistent wall thickness at every join and side arm with domed ends to sealed off tubes, rounded shoulders leading to solid well run in button seals, supported internal seals, inverted and side seals. When making glassware for use under vacuum or pressure, work to the highest standard you possibly can while avoiding flat bottomed flasks and tubes where possible.

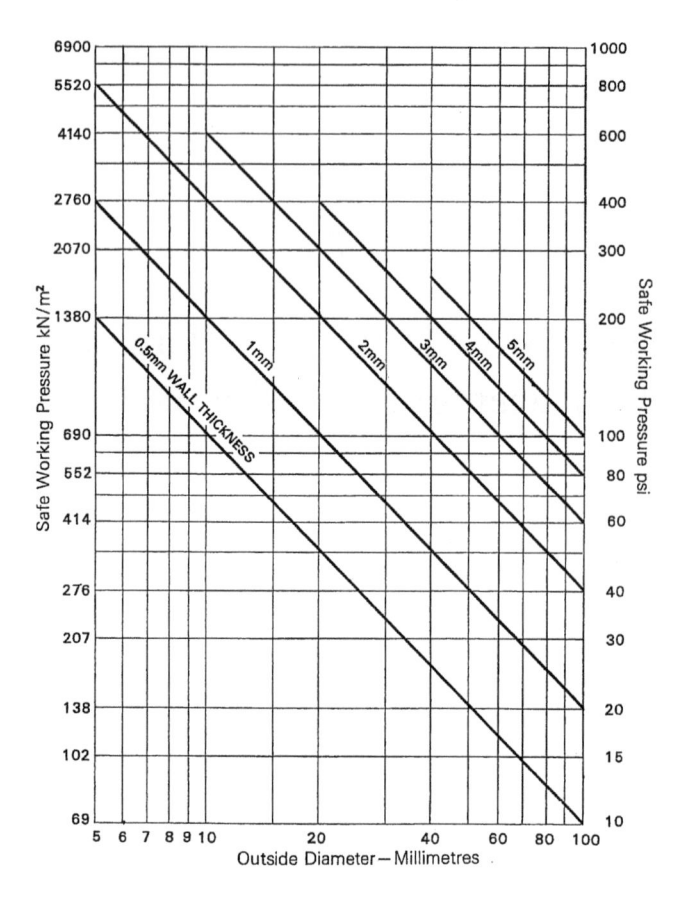

Fig. 7.1.

The exception being the "Buchner" flask which by its very design and wall thickness will withstand daily usage under vacuum. If you have the opportunity, look at the bottom of any Buchner flask that has been in use — there may be and often are scratches (Fig. 7.2). Recommend that the flask is renewed and that when in use that the base is rested upon a layer of softer material rather than directly on hard surfaces.

Vacuum Measurement — A Story of Time and Place

If one stands at Greenwich, London, UK from where the Prime Meridian is measured at 0° longitude, the weight of the worlds atmosphere amounts to

Fig. 7.2.

14.7 pounds per square inch or 1033 g per cm². This atmosphere will support a column of mercury 29.9 inches or 760 mm and a column of water 33.9 feet high.

The 760 mm column of mercury is significant — if divided into 760 units called Torr after Torricelli, the man who realised its importance in pressure measurement.

$$1 \text{ Torr} = 1 \text{ mm of mercury.}$$
$$760 \text{ Torr} = 1 \text{ atmosphere.}$$

Any fraction of an atmosphere is a partial vacuum and is a negative pressure. The metric standard system is based on unit force i.e. Newton per square metre and the name of the unit is the Pascal.

The standard atmosphere is 101,325 Pascal. Further subdivision means that 100 Pascal is equal to 1 millibar.

1 millibar equates to $^1/_{1000}$ of a bar. The standard atmosphere is equal to 1013.25 millibar.

$$1 \text{ millitorr is } ^1/_{1000} \text{ of a Torr.}$$
$$1 \text{ millibar is equal to } 100 \text{ pascal, confused?}$$

Though there is standardisation through the International Organization for Standardization (ISO), vacuum measurement within a laboratory setting can seem to be in a state of flux where two adjacent laboratories may be taking readings on entirely different scales, at high vacuum the difference is infinitesimal, so long as the scientist is constant in the use of one system or the other — conversion factors abound to bridge the divide between Pascal, millibar, Torr, millitorr.

There are a wide range of methods for measuring degrees of vacuum from the mercury U tube manometer connected directly to both the system and atmosphere to give a direct reading.

The McLeod gauge also uses mercury in glass, the design of which traps a given volume of gas in relation to the degree of vacuum present on either a direct or logarithmic scale. This is an accurate system between 10^{-4} Torr and 10^{-6} Torr but can be affected by the presence of water vapour and oil vapour.

The Bourdon gauge uses a sealed metal tube with a mechanical lever connection to indicate the deflection caused by the internal vacuum force, some Bourdon gauges are fitted with a transducer to turn the movement into an electronic reading.

Pirani gauges use a heated platinum wire, the temperature of which will increase in relation to the higher vacuum. The resistance in the wire can be measured and the vacuum indicated.

The useful range is from 0.6 Torr to 10^{-4} Torr.

The most sensitive ionisation gauges can read down to 1×10^{-12} Torr.

The degree of vacuum in laboratory condition are as follows:

- Low vacuum 1×10^5 to 3×10^3 Torr,
- Medium 3×10^3 to 1×10^{-1} Torr,
- High 1×10^{-1} to 1×10^{-7} Torr,
- Ultra-high 1×10^{-7} to 1×10^{-12} Torr.

The means of producing the vacuum itself ranges from the water jet ejector which will give a sufficiently reduced pressure to operate a Buchner filter unit.

The mechanical rotary pump with either one or two stages will give a reliable means of displacing gas to a much greater degree, so much so that mechanical rotary displacement pumps act as a backing pump to create sufficient low pressure to enable an oil or mercury diffusion pump to work. The eventual limitations of this type of pump become apparent when they have reached their ultimate vacuum when the very operating liquids within start to effect the vacuum with their own vapour pressure.

The highest vacuums produced in the laboratory are currently by high speed turbo molecular pumps running on magnetic levitating bearings to 1×10^{-12} Torr.

All of these various systems have their own specific operating procedures which in turn generate various levels of vacuum. Many years ago before becoming a Scientific Glassblower, I was training as an industrial chemist, part of the training was surface area determination on one of the products that the company made. This was done on what I now know to have been a vacuum line complete with a mechanical rotary vacuum pump or backing pump used to support a three-stage mercury diffusion pump, two liquid nitrogen traps which led to a manifold with a series of greased high vacuum taps. There was also a large mercury filled Tourpler pump and finally a series of tall mercury manometers.

I can remember it in my mind's eye but cannot for the life of me remember the operating sequence. I do remember watching the in-house glassblower repairing it with awe and wonderment. 50 years on, after some years training I have made a number of vacuum lines myself. In more recent years, I have been asked more and more to make Schlenk lines which are interconnected double manifolds where one manifold is charged with an inert gas, the other a vacuum line operating in the region of 10^{-2} to 10^{-4} Torr.

The line does not require a diffusion pump but uses only a mechanical rotary vacuum pump. There are one or two traps dependent upon the systems ultimate use. The inert gas line is supplied with either nitrogen or occasionally argon, the choice of which is dependent on the chemistry being conducted.

A notable feature of the inert line is a pressure relief point, usually an oil filled bubbler which has over time become the relief system of choice because when filled with nitrogen the rate of gas supplied can be regulated simply by counting the bubbles, which at a glance gives an instant confirmation that there is a supply of inert gas as the chemistry used as a Schlenk line is for air sensitive compounds.

The Schlenk line allows this type of chemistry to be conducted by switching from the inert to the vacuum line without the need to disconnect from a single manifold line, to then do further chemistry. The two manifolds are linked by either double oblique greased high vacuum taps (Fig. 7.3) or by

Fig. 7.3.

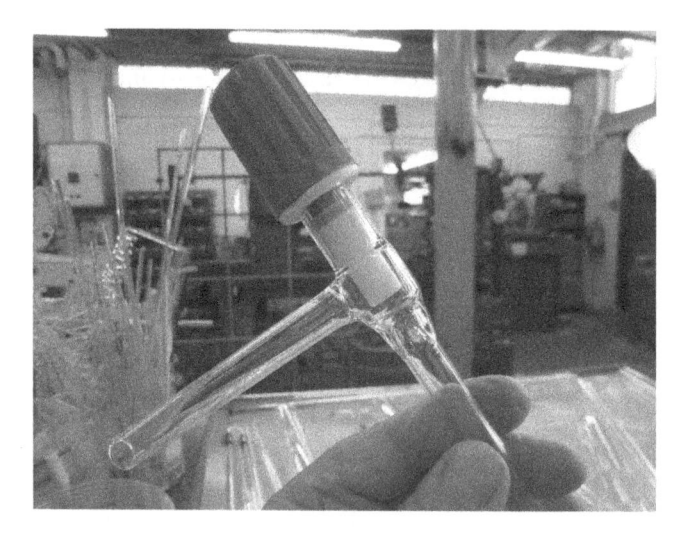

Fig. 7.4.

grease free PTFE plunger type taps (Fig. 7.4). Schlenk vessels are linked to the manifold via a flexible tube.

The flexible connection is made via a push fit to both the manifold and flask, whereas a true vacuum line operating at between 10^{-6} Torr and 10^{-8} Torr

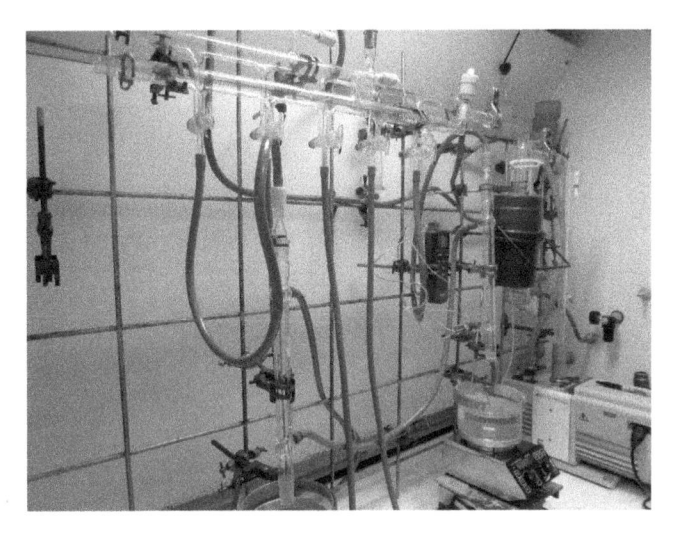

Fig. 7.5.

would use a joint to maintain the integrity of the vacuum. The flexible connection should not be too long that the fume cupboard (Fig. 7.5) in which the Schlenk line has to be located does not become cluttered with snake-like plastic tubing or too short that the Schlenk flasks are suspended from the manifold.

The operations carried out on the Schlenk line range from injecting liquids through septums into nitrogen filled vessels, counter flow techniques, distillation and even filtration in an inert atmosphere. The inline trap can be used to collect distillate but its prime objective is to protect the rotary vacuum pump. To cool the trap, either solid carbon dioxide or liquid nitrogen is used (Fig. 7.6).

A point of caution: When using liquid nitrogen, the low temperature can condense oxygen from air if the tap on the pump side leaks or is left open. The oxygen reacts violently with any organic matter.

The responsibility of the glassblower in making a Schlenk line is to ensure that there are no leaks or pin holes that might entrain air and oxygen into the system.

It will destroy air sensitive components.

Liquid oxygen can by its very natural property destroy a fume cupboard, destroy a building and destroy people!

Fig. 7.6.

Points to Note When Making a Schlenk Line

The Schlenk line consists of two parallel manifolds with the nitrogen line to the rear. The ends of the manifolds are best fitted with joints to enable easier cleaning. Make up the manifolds and mark the positions of each set of taps. If using double oblique greased vacuum taps, the arms will need to be adjusted to allow for the distance between each manifold. When using greaseless PTFE plunger taps, think carefully about the orientation and make up each pair of taps. "Think and measure twice as much as do" at this stage (Fig. 7.7). Clamp or use a jig to hold both manifolds in position, seal one end, attach a blowing tube to the opposite end, blow a hole at the marked position to one end, remove the blowing tube and seal with a cork, repeat on the second manifold, cut the arms of the tap to length and check the alignment.

Use a hand torch to make the two holes seal and run the joints in, this first operation is perhaps the most difficult, the remaining joins are just awkward. Check the alignment of the tap set at 90º to the manifold, flame anneal the first two joins and proceed along the manifold using the previous tap to align each stage until complete — remove from the jig and check the verticals once more, then anneal the manifold.

Fig. 7.7.

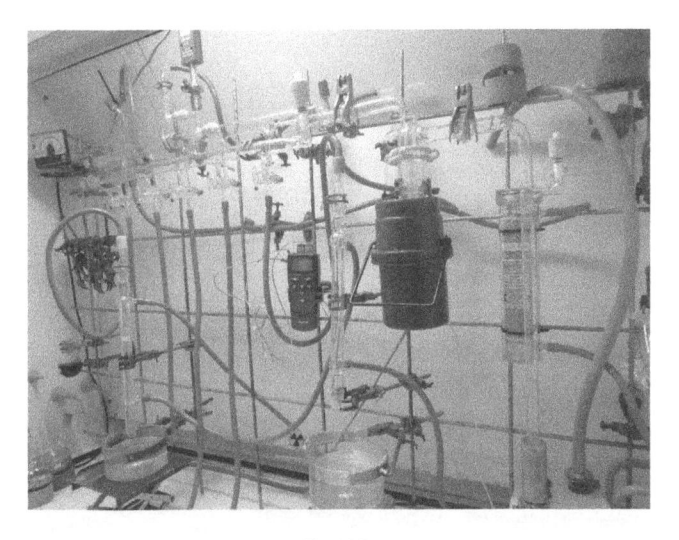

Fig. 7.8.

There will be further additions to be made for the vacuum gauge and oil bubbler via sockets which can be put in place once the glass had been annealed and the anxiety of making a series of double joints is over (Fig. 7.8). The trap head is a supported seal with a side arm. The sequence is similar to the making of the much smaller Dreshel head (Fig. 7.9). The trap body is a

Fig. 7.9.

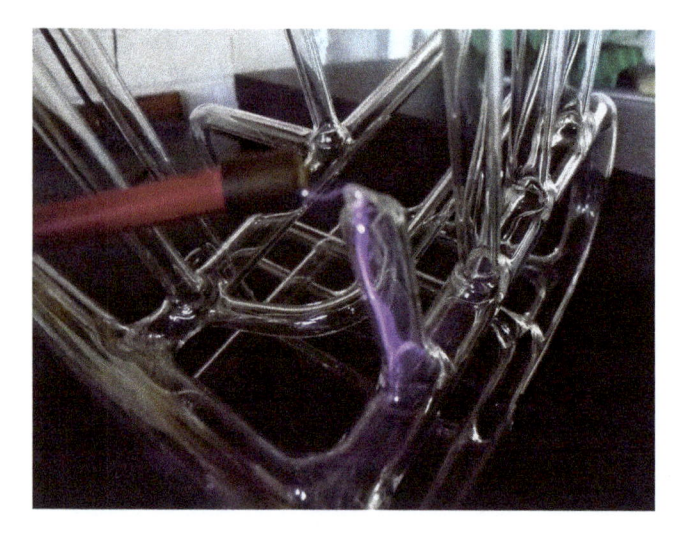

Fig. 7.10.

straight join to the correct size socket with a rounded end. The bubbler can be a standard Dreshel head in a jointed Dreshel flask.

A similar design fully sealed can be made bearing in mind that the level of oil in the bubbler will affect the pressure in the manifold. There are a variety of designs and many variables reflecting the final usage of the glassware.

Discuss with the chemist their requirements, desires and wants!

From your point of view as a Scientific Glassblower, ask questions to fully understand what is required and advise accordingly.

Is the design practicable? Are all the raw materials available? Is there sufficient flexibility within the design dependent on the length of the line and would the inclusion of a spherical joint be advisable? Has a valve been set in the line between the pump and the trap to return the system to atmosphere after work has been completed? Will you remind the new inexperienced chemist about avoiding trapping liquid oxygen?

Finally, have you assembled and evacuated the complete system and sought out any leaks using a "Tesla" coil? (Fig. 7.10)

8
Silvering of Glassware

Strip Silvering

First, allow me to explain that on leaving school my first job was to work for a company that made sodium silicate.

Sodium silicate was made from sand and soda ash, that is, SiO_2 and Na_2CO_3, the common name is water glass. Sodium silicate was made in large "Siemens" regenerative glass furnaces; it was a continuous process producing two tons of glass per hour, seven days a week. The glass produced was dissolved in water to produce a liquid of various specific gravity. As a trainee, I was allowed to use a Hydrometer and measuring cylinder to test the solutions.

Once a year a batch of potassium silicate was made which resulted in the amazing sight of five tonnes of molten glass pouring from the furnace into a shallow iron tray where a lake of molten glass sparkled as it cooled into a five tonne crystal. I was moved into a laboratory that analysed further sub-products such as alumina silicate, sodium meta-silicate and eventually soap-based detergents using sodium silicate and sodium perborate.

I was occasionally allowed out onto a pilot plant to assist, this was interesting applied chemistry but always had to return to the analytical laboratory which I found to be a brain numbing experience, as this was an age when routine analysis was done by wet chemistry techniques.

The products were made continuously so analytical control had to be 24/7. The thought of knowing that the following day I would know exactly the same analysis was not an existence I would savour. There was a positive aspect to this period in that I came into contact with laboratory glassware, which I now know was a vast range from specific gravity bottles to vacuum lines complete with glass diffusion pumps, mercury in glass manometers and Toepler pumps plus all the various vacuum taps and associated paraphernalia.

The more I used the glassware, the more fascinated I became in learning how it was made, fortunately there was a glassblower on site. I didn't have to go far to ask the questions. Not all laboratory work is fascinating, a point in case was the chloride test which involved mixing the sample solutions with a solution of silver nitrate, if chloride is present the result is instant as silver chloride precipitates from solution. I was twenty years old. At the time of writing this — 50 years of Scientific Glassblowing in a research environment was where I have wanted to be and have been fortunate in attaining that aspiration.

There were glimmers of knowledge from my time in the laboratory and it is called experience. Experience in handling chemicals in a safe manner within the laboratory environment. Soon after finishing my initial training as a Scientific Glassblower, I was asked to make up silvering solutions from the raw materials using the "Brasher" method which amounted to making two solutions, one a solution of silver nitrate, the other a 'reducing' solution which when mixed together pure silver comes out of solution which adheres to any glassware it comes in contact with.

It was and still is important to wash every item the solutions come in contact with in distilled water or deionised water because: if one uses or washes the glassware with tap water, silver chloride will come out of solution rendering the silvering solutions useless.

Rule 1 — Wash all glassware that will come into contact with the silvering solutions with distilled or deionised water. This means the equipment and the jacketed vessel or Dewar. In the instructions, it will say "do not store made up solutions overnight". One may be tempted if you have silvering solutions left over and more works to do the following day. Read the following true experience and make your own mind up as to whether to obey instructions.

I worked for a time with a gentleman glassblower by the name of W. G. C. Hunt who described a time when he had made up some silvering solutions and had left the excess overnight as he intended to continue silvering the following day. He noticed that there were small black solids in the bottom of the beaker, he took a glass rod and started to stir the solution, when the solids refused to go back into solution he crushed one with the end of the glass rod. The beaker plus the side of the fume cupboard disappeared instantly. The black solids were silver fulminate — highly explosive!

Fortunately, Mr W. G. C. Hunt was unhurt, although as he turned around to his equally shocked colleagues he took off his eye protection goggles only to reveal his white eyelids and rapidly darkening face — within a few moments he looked like a PANDA in negative.

Rule 2 — Never store silvering solutions overnight.

Rule 3 — Always use your personal protection safety equipment when using silvering solutions.

Rule 4 — Make up the silvering solutions in a fume cupboard "sufficient unto the day".

Silver nitrate when in contact with the skin will discolour it, so use NITRILE gloves when making up solutions. Modern silvering solutions do not use the "Brasher" method, solutions are premixed and utilise a different formula so there is less chance of detonation but the above four rules still apply.

Anneal the glass to be silvered, wash the annular space with either distilled water or deionised water, make up the sensitising solution and wash the annual space with this solution (Fig. 8.1). Take the silvering and reducing

Fig. 8.1.

Fig. 8.2.

solutions which have been diluted to the manufacturer's recommendations, premix the two solutions in a beaker and pour immediately into the annular space. The solutions should darken and silver should be seen coming out of solution depending upon time and temperature. The process is finished when the solution changes to a muddy brown and an "even" silver coating is deposited (Fig. 8.2); flush away the used residue using distilled water. A second of third coating can be added, if necessary.

After washing, allow the glassware to drain then dry in a warm oven at 150°C — do not re-anneal as annealing temperatures are far too high which will cause the silver to tarnish.

The inlet to the vessel must be of a bore size that will enable sufficient silvering solution to enter quickly. When there is a narrow tail or constriction through which the silvering solutions have to pass, the vessel can be evacuated and the solutions introduced via a "T" bore tap.

Strip silvering can utilise the lathe or engineers' 'V' block, the essential point to note is that the glassware should be held horizontally thus I prefer to use the lathe which I know to be true. The glassware is annealed, washed, sensitised and dried. Two opposing inlets/outlets will have to be made on the vessel to allow access for the two stage silvering process (Fig. 8.3).

The premixed solutions are introduced gradually from underneath via a tap so as to avoid ripples to a pre-determined point on the glassware. The spent solutions are drained away through the same entrance. The glass is

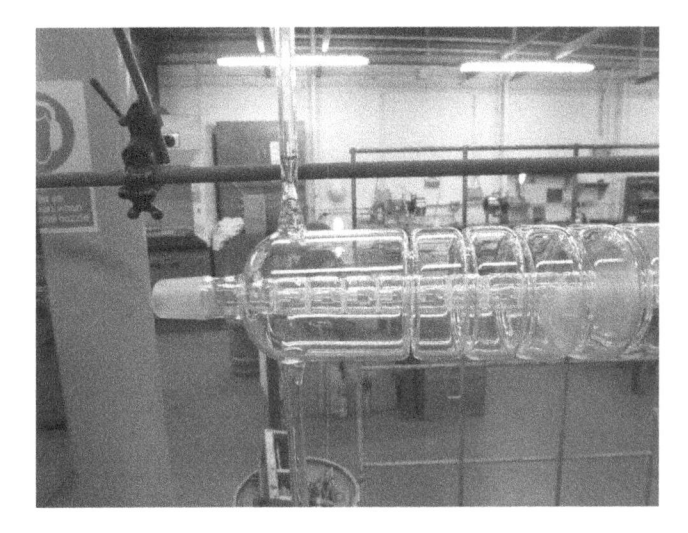

Fig. 8.3.

Fig. 8.4.

washed and dried before returning to the lathe and aligned once more. The final level marked and the solutions introduced from underneath once more. The glassware is then drained, washed and dried. One arm is sealed and the second constricted, then attached to a vacuum system for outgassing, evacuating and sealing off (Fig. 8.4).

9
Packed Columns

Many years before the development of fractional distillation columns, simple distillation took place in glass retorts which consisted of a flask with a jointed entry port and an air condenser protruding from the flask at an angle. Simple separation of mixtures using heat where the lowest boiling point fraction vaporises first then as the vapours pass along the air condenser returns to the liquid phase as a purer distillate (Fig. 9.1).

Simple as this is, retorts were being made in Venice since the fifteenth century, used by alchemists and the early chemists over many years. With the development of the Liebig condenser in the nineteenth century, using a cooled jacketed vessel enhanced the distillation process especially when this type of condenser was used vertically which led to the development of the fractionating column, which enabled fractions with closer boiling points to be separated one from another. The packing within the column became more complex, offering far greater options to suit a given chemical type. The packed fractionating column does not work in isolation but must be part of a system consisting of the following:

A boiling vessel, the column which is often jacketed and strip silvered; the fraction take off and receiver, all topped off with an efficient condenser.

Temperature control is deemed to be critical at every stage.

The design of the column including length, diameter and type of packing corresponds to a figure called the theoretical plate which determines the column efficiency. At this point I will quote verbatim the response received from

Fig. 9.1.

a design engineer as to his requirements, considerations and calculations which affect directly the overall height and diameter of the column.

Information Regarding the Use of Such Distillation Columns

We would consider operating conditions such as temperature, pressure and material compatibility before selecting a suitable distillation column.

We often use PRO-PAK Monel random packing which can be obtained as 0.16″ sections of metal ribbon. This suits our narrow distillation columns as, typically, this size of the packing should be less than 1/10th the diameter of the column. If the packed height is greater than 1 m, then channelling can occur, whereby the liquid flow is forced towards the walls, resulting in reduced efficiency. Inter-distributors can be fitted every 1m to redistribute the liquid across the column cross-sectional area. The column diameter should be determined by the vapour flow rate required to achieve the rate of processing and degree of separation. As is often the case, in reality, columns are selected based on availability and are made to fit the duty. Column packing can be operated up to its flooding rate, which provides the maximum vapour velocity flux as a function of the packing employed and fluid densities (liquid and vapour). Beyond this point, the descending liquid becomes entrained by the rising vapour. The pressure drop across the column is not normally an issue unless the distillation must be conducted at very low absolute pressures. Column performance is measured in terms of the height of an equivalent theoretical plate (HETP). The HETP achieved is a function of the packing employed, fluid physical properties and operating conditions. The PRO-PAK 0.16" mentioned above can achieve HETP values in the region of 1–2".

More recently, we have trialled the use of structured packing which has the advantage of higher efficiency (reduced HETP) but is more prone to fouling.

Recently, I was asked to make and install two columns, the overall height of which were 2.4 m and 3.67 m. The jacketed boiling vessel was embedded within a solid outer box construction and was set in place using a spirit level and lasers to confirm that the column was truly vertical, as each component was added, its weight put the glassware into compression.

Glass like most liquids is strong in compression and weak in tension so having a substantial and solid base each section was added with confidence. The columns were supported within a metal frame for lateral stability and protection from the occasional clumsy chemist. Both columns were constructed within a high pressure blast chamber.

The engineers and chemists did for a time consider using an Oldenshaw all glass bubble plate column, settling instead for the packed column as this gave a greater number of "theoretical plates".

The final design for the column consisted of a strip silvered jacketed column fitted with the required size joints — top and bottom; there were also a series of bellows in the outer jacket to make allowance for any expansion due to temperature changes.

The fraction take off was also jacketed and fitted with a solenoid operated swinging bucket. There were certified thermometers in combination with electronic temperature measuring devices throughout the system as the temperature readings were critical and had to be precise. At the very top of the columns were large triple coil condensers. Both columns were a joy to make. To see them installed, working and giving results justified all the hours of practice and the faith shown to me by my tutors — to whom I offer my sincere thanks.

10

Round Bottomed Flasks — Various Techniques for Attaching Side Arms

To begin, I will describe the method that I was taught many years ago as to how to attach a variety of necks to a flask at the bench followed by the production line techniques using a vertical jig and hand torch and finally the use of a lathe to hold and attach joints to larger flasks.

You will notice that "standard" three neck flasks have their vertical joints positioned within the overall diameter of the flask and that the angled necks are directed to the centre at the bottom of the flask. I have worked for many years in a research environment where my customers have often used three little words which were "can you just?"

Can you just alter the angle of the joints so that the reentrants to the flask are not fouled by the stirrer which rotates in the centre of the flask? I have worked for many years with Organic Chemists who often ask for three different size joints to be added to a standard flask:-

One vertical, one opposite at an angle and a third positioned centrally with an angle betwixt the two. One will find that research chemists will use standard catalogue stock items but are delighted when someone can answer "yes" to the question, can you just?

For the exercise, I have taken an arbitrary size flask and joints. Use the method to suit your needs, skill level and material availability to the extent of using straight tubes with which to practise.

Fig. 10.1.

Exercise — Materials Needed

500 mL round bottomed flask with
    B34 socket centre neck
    B24 vertical side socket
    B19 angled socket opposite
    B14 angled socket mid-point between the B24 and B19.
    Refer to photograph to confirm the configuration (Fig. 10.1).

Preparation of the Joints

Cut the B19 socket by either the hot point method or cut the disc 3 cm from the base of the socket. Cut the B24 socket at an angle of 40° on the diamond cutting wheel using a suitable support starting 3 cm from the base of the socket. Cut the B14 socket at 40° angle 2 cm from the base of the socket. Wash the joints in clean water to remove the cutting debris. It is advisable to cut spare side arms of each size. Although this may prove to be unnecessary after some years of practice, it is always best to cover all eventualities as occasionally joints crack when being warmed in preparation to starting the join.

    Take a B19, B24 and B14 cone to use as a handle, wrap a single thickness of corrugated graphite tape around each cone and insert into the prepared sockets (Fig. 10.2(a)). The holder for the central B34 socket consists of a B34

Round Bottomed Flasks — Various Techniques for Attaching Side Arms 359

Fig. 10.2.

cone joined via an unequal join to a 30 cm length of 12 mm ø medium walled tubing, the end of which is flame polished. Using corrugated graphite tape, fit the B34 cone into the flask socket and tap the base of the holder onto the bench to ensure that the joint is firm (Fig. 10.2(b)).

Three corks will be required to seal off the sockets after joining to the flask. The large end of the cork should make the seal near to the top of the socket (it makes it easier to remove). Finally, take a 30 cm length of 12 mm ø medium walled tubing flair one end to 15 mm ø. Select a suitable cork with which to seal the tube later in the process.

We are about to start!

First Method

Place the flask with its holder in the left hand and rotate to ensure that the left hand grip is correct, rest the elbow on the bench, bend the wrist, support the glass with the second, third and fourth fingers and rotate the glass using the first finger and thumb. Adjust the burner to produce a feather-like flame which is aimed at the base of the flask until specs of sodium flames appear. Whenever possible do not allow the flame to play directly on the ground section of any of the joints. Reduce the size of the flame to give gas, air, oxygen flame, continue to aim the flame at the centre of the base of the flask until it softens, remove from the flame, blow to produce a dome return to the flame aim the flame at the top of the dome until it melts, withdraw and blow to produce a thin bubble which is sliced off with the glass knife to reveal a hole. Seal the B34 socket holder with the cork and make a straight join between the base of the flask and flared end of the 12 mm ø tubing.

Run the join in — the left hand will be dominant especially with the momentum caused by the weight and size of the flask. The right hand has a much lighter touch as it follows the action of the left. Once the initial join has been run together, withdraw from the flame and rotate as you pause before blowing up the join while holding the glass vertically (Fig. 10.3). Rotate to centralise.

Warm the flask in a feather-like flame to disperse any strain. Allow to cool slightly then place the 12 mm ø tube into the left hand, remove the cork from the B34 holder and place in the left hand stem. We will always work from left to right with the weight in the left hand. Warm the flask in a feather-like flame and heat the shoulder of the flask, be aware of the edge of the flame near the B34 socket at all times. Once the shoulder is warmed, reduce the size of the flame and heat a point on the shoulder where the joint will lie, blow to form a dome. One has the option at this point to either blow out the dome and form a thin bubble which will be sliced off with the glass knife or keep

Fig. 10.3.

Fig. 10.4.

the dome in contact with the flame while blowing so that the "blow out" produces a "melt back" hole without loss of wall thickness from the flask.

Reheat the hole and use a carbon rod to elongate the hole to reflect the shape and size of the B24 (Fig. 10.4) socket cut edge. Take the B24 in the right hand and heat the edges of both the joint and flask after increasing

Fig. 10.5.

the intensity of the flame. Once the two edges are hot, join the edges together–touch–pull–blow. Assuming all is well and the seal made, we are now about to "run in" the join, we are in fact making a "straight join". Start by adjusting the flame to create a long thin intense gas, oxygen flame. Aim at the flask with the right hand side of the flame in contact with the join, watch as the glass melts within a narrow band as surface tension forms a neat radius between the joint and flask (Fig. 10.5).

Little blowing, if any, is required as you allow gravity to form the radius. Once the inner part of the join has been completed, adjust the burner to give a feather-like flame to balance the temperature as the intense flame will create strain, watch for the sodium flame and when confident that the temperature is balanced reduce the size of the flame once more and concentrate on running in the remaining initial join in the two sections along the sides, as you melt the glass, hold the flask and join horizontally allowing gravity to retain the diameter. At this point, warm off the flask with the "feather" flame. Reduce the size of the flame and create a circular hole opposite the first, use the carbon rod to open the hole to the same diameter as the B19 socket — heat both edges in a flame of similar width to the joint. Take the B19, get the edges hot–touch–pull–blow.

Reduce the size of the flame and run the join in using the thin oxygen gas flame while holding the flask so that the join is affected by gravity and shaped by surface tension. Warm up the flask with the feather-like flame. Blow a hole between the two other joints with the B24 positioned to the right. Elongate the hole, make the initial join and run in. Warm off the flask. Remove the small cork from the holder, remove the joints and replace with

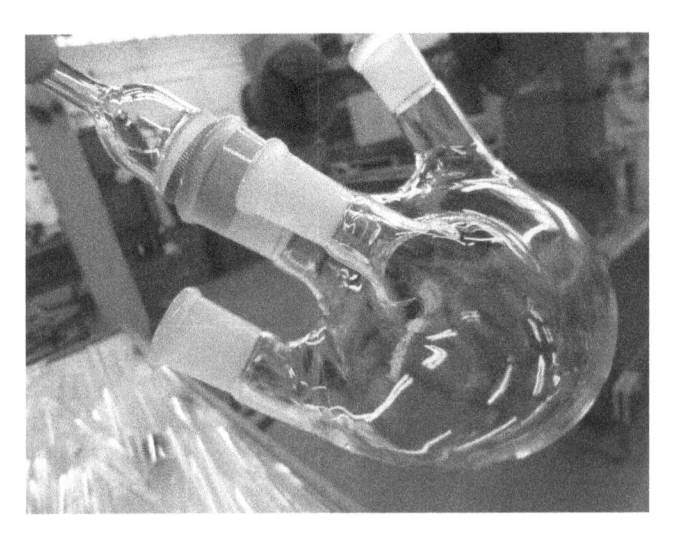

Fig. 10.6.

the corks, using a narrow flame burn off the holder from the base of the flask, increase the size of the flame then heat and blow the base to return the radius to the base of the flask. Warm the flask in a feather-like flame. Adjust the flame to give a noisy gas air oxygen flame, incline the flask a little and wipe the flame from the stem of the joint to the flask, gradually the stem and flask will merge together to give a smooth transition between the two. Now is the time to ensure that the angles of each joint are in accordance with the customers wishes (Fig. 10.6).

Practice and the use of the method will reap its own rewards in time. Using this method, one will be able to add numerous joints at various angles to flasks from 50 mL to: one, two, three litre flasks and vessels. One may be concerned about the weight as was my apprentice who was 6 ft 8″ tall and as strong as a very strong thing! He complained about the weight of a small vessel he was adding joints to, I immediately took the flask and weighed it — 36 grams! I did not bother to explain knowing in time he would realise it was the weight of his own arms he was fighting against — I watched as he readjusted his posture and learned one more little lesson. As scientific glassblowers we are not handling exceptional weights but we are controlling molten glass for quite long periods of time. The control of molten glass has to be total — failure to concentrate and act accordingly means that gravity has won.

Fig. 10.7.

Second Method

Proceed as per previous method to the point at which the initial join has been made, instead of running the join in at the bench I would like for you to run the joint in using a hand torch. Once the initial join has been made, we will hold the flask in such a way that when the join is run in it does so by gravity. Use a narrow fierce flame from underneath the join — heat between a third and a quarter of the join aiming predominantly at the flask side of the join and watch as the "liquid" glass runs the two together, as soon as the initial ridge disappears — STOP — and hold the glass stationary for a while as the join cools, continue working around the seal until satisfied all is run in (Fig. 10.7). This is a particularly quick and pleasing method for making the join especially when making a straight join to a flask, little or no blowing is required, so that once the initial join has been made the joint holder can be removed.

Third Method

Uses a combination of the previous methods but the flask or extended tail is held in the left hand chuck of a lathe. A dog leg jointed holder of various joint sizes is held in the right hand chuck so that the socket aligns with the

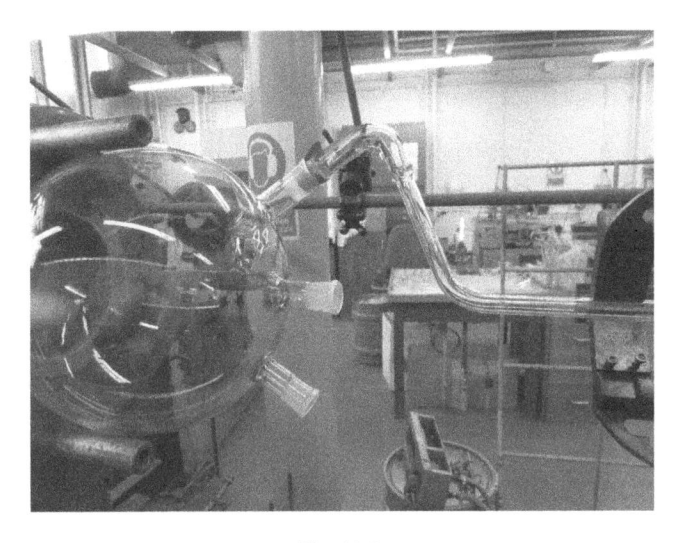

Fig. 10.8.

flask at the required angle, a hole blown at the proposed point of contact. The two edges are heated and the joint brought together with the flask and run in without fear or losing control of the join. It is important to prewarm the flask then during and after each join has been made. This method is designed to be used when working with larger, heavier vessels (Fig. 10.8).

Fourth Method

On a production line, conformity to a standard size and shape is paramount. The joints have to be held in a jig so as to eliminate variations from the standard. The flask is often held vertically while the joints are held at a predetermined angle in relation to the size of the flask. The joints are attached with a torch fitted with a semi-circular set of jets. As the join is being made vertically, the glass will flow from the stem of the joint into the flask ensuring no loss of wall thickness.

Take a close look at a standard flask, notice how well the join has been run in without bubbles, join line or uneven wall thickness. When you work at the bench free hand, use a lathe to take the weight or a jig to support the join.

This is the minimum standard expected.

11

Standards of Competence

The definition of a standard is as follows:

A measure of extent, quality, value etc. established by law or custom as an example or criterion for others — the degree of excellence required for a particular purpose and recognised as a standard for imitation or comparison.

The standards expected of a student have many facets, from the mentor or trainer, the employer, the individual him or herself, the customer and standards imposed externally by International Organization for Standardization (ISO) and professional bodies.

The standards attained through recognised exams which are set and graduated by Glassblowing Societies and the constant assessment by one's mentor or trainer through pre-set expected attainment levels will reassure and guide the students' expectation.

For example, as a guide only, adapting the following to suit individual circumstances, a training schedule for a full time trainee Scientific Glassblower is given below.

Period Covered — 6 months

1. Knowledge of the material.
   Subjects to be covered: Chemical compositions, working, strain and annealing points.
   Temperature limitations, storage and handling, technical info, etc.
   Materials — Soda lime, Borosilicate and Quartz (fused silica).

2. Hand manipulation

   Cold working

   Correct and safe methods of cutting flat and tubular glass, size limitations for cutting by hand, care and maintenance of cutting tools, protective equipment.

3. Burners (Bench)

   An appreciation of the following:

   Fuel gasses used, cylinder and regulator safety procedures, non-return valve requirements

   Temperature relating to different gasses/glasses,

   Subjects 1 to 3 to be covered over a four-week period.

4. Hand manipulation (Bench)

      Preliminary hand working techniques

      Rotation and grip

      Pulling spear points

      Subject 4 to be covered over a four-week period. Time taken at this stage 2 months.

5. Hand manipulation (Bench)

   Utilising spear-points

   Straight joins

   Rounding off

   Unequal joins

   T pieces

   Cross pieces

   Subject 5 to be covered over a four-week period. Time taken at this stage 3 months.

6. Hand manipulation (Bench)

   Utilising spear — points

   Bulb blowing — concentricity and size

   Thistle funnels — unequal joins and reaming

   Subject 6 to be covered over a two-week period

7. Hand manipulation (Bench)

   Utilising spear — points

   Small diameter bends — 90°

U and angular Y joints

Button seals

Small trap construction

Subject 7 to be covered over a two-week period. Time taken at this stage 4 months.

8. Hand manipulation (Bench)

    Small apparatus

    Spray

    Atomiser

    Subject 8 to be covered over a one-week period.

9. Hand manipulation (Bench)

    Larger bends — 90°, U and angular Y joins

    Subject 9 to be covered over a one-week period

10. Hand manipulation (Bench Burner)

    Splash heads

    Water pump

    Small trap

    Soxhlet extractor

    Viscometers

    Subject 10 to be covered over a six-week period. Time taken at this stage 6 months.

Standards Maintained by: The Individual and Expected by the Customer

Be aware that your standards become obvious in the work you do — sometime in the future you will recognise your own work and you will remember how you felt on the day you made it.

Slip shod work will reflect on you, so aim for the highest standard you can attain on the day given your level of experience and training and accept that people like me will make allowance for the length of time you have trained, they will understand the amount of effort needed to attain and maintain the standard you have reached so far.

One will become more and more aware of the initials ISO affecting every aspect of glassblowing and engineering standards from the interchangeability of socket and standard 1:10 tapers to the final ground surface, all aspects from the raw material to final product will be covered by the International Organization for Standardization which is a non-governmental international organisation with a membership of 162 National Standards bodies such as the British Standards Committee in 1901 and the American National Standards Institute to name but two.

The ISO is based in Geneva, Switzerland.

Standards of workmanship may be self-imposed whilst being guided by a mentor. While National Glassblowing Societies such as the British Society of Scientific Glassblowers through its Board of examiners set competency exams at various levels of a trainees development.

The Parable of a Trainee Glassblower

While still a trainee glassblower, a piece of glassware was requested by a chemistry student who I must admit I was not fond of.

When he came to collect the glassware on the agreed day I realised that I had not completed the work and tried to cover up the fact by quoting difficulties of supply and anything else I could think of.

After he had left, the glassblower who was teaching me and who had been listening gave me the most intense dressing down I have ever had in my adult life.

The lesson was learnt. Ever since that day and without fail I have shown respect to the customer whoever they may be with a professionalism as befits the service I provide.

Since that day I have learned that in time a trust is earnt which in turn may even develop into respect.

If one sets out to provide a service, it is sometimes difficult to say "No it cannot be done in glass" so long as you have thought it through after understanding the customers' requirements and have related them to your own level of experience, the tools and equipment available, the limitations of the glass in relation to chemical reactivity or temperature and strain constraints.

Finally having offered alternative designs, then and only then will you be sure that the word "NO" is based on firm knowledge.

I have found that the customer can and will respect the advice.

12

Drawing on Experience

12.1 The Manufacture of a Quartz UV High Temperature Coil Photoreactor

Michael Baumbach

H. Baumbach & Co., UK

The manufacture of a Quartz UV high temperature coil photoreactor was chosen as it encompasses many of the desired qualities that Quartz exclusively brings to scientific experiments very, for example, high temperature upto 1200°C and Light transmission into Ultraviolet (UV).

Quartz must be cleaned to exacting levels as contamination in clear Quartz can lead to devitrification (recrystallising of the Quartz structure causing the glass to turn white and flake). This will lead to a weakening of the final structure, where Quartz flakes enter into the experiment and lead to discolouration of the material itself, ultimately causing a 'fail in the final quality check'.

The photoreactor consists of four parts. The inner tube, the coil, the outer tube and the two side arms.

Firstly, the correct material must be selected so that both dimensions and UV requirements are met. The inner tube is made first, a simple test tube with a flared end. The end must match dimensionally with the outer diameter of the outer tube. Once completed, set aside and keep clean. The outer tube is made next, again a test tube is created keeping with the dimensions required. A very small tube is then attached to the rounded end. This will help avoid fusing problems later. Next is the coil, this is the fun part, taking a mandrill (slightly larger than the outside diameter of the inner tube) the required tube is wrapped around the mandrill so the correct number of turns are kept. Special attention must be paid to avoid the tubing from touching as this can lead to the tube cracking or breaking when it cools down.

Once the desired number of turns is completed, leave the coil to cool down and remove from the mandrill. When removed, the ends must be shaped to touch the inner wall of the main outer tube.

Place the coil over the inner tube and mark the exact location the side arms need to be, then fuse the coil directly to the inside of the outer tube, making sure that the coils still allow the inner tube to sit comfortably. Now fuse the flanged section of the inner tube to the outer tube. Allow to cool then fuse the side arms into position.

Anneal the Quartz making sure it is still meticulously clean. The final process is to remove the small tube which is located on the outer tube. Heat and pull away, making sure to avoid thinning the end.

Drawing on Experience 373

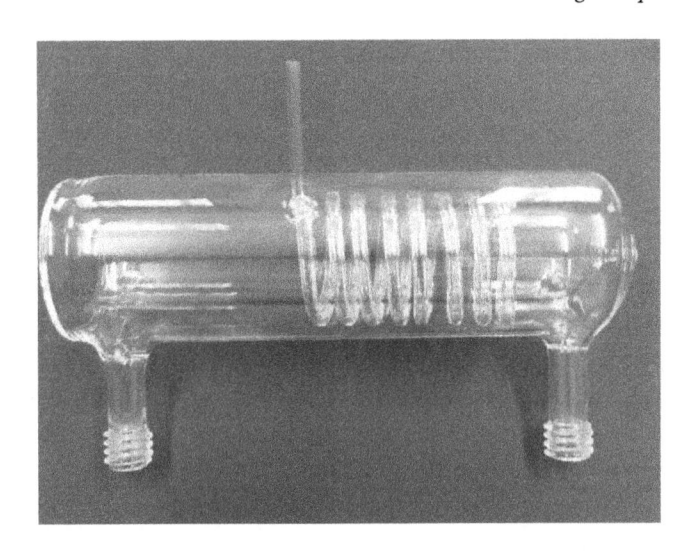

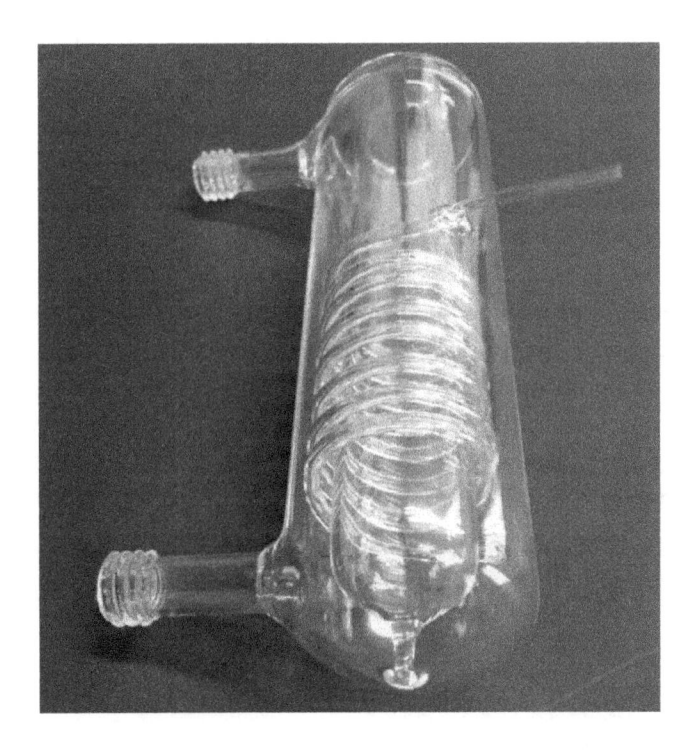

12.2 Laser Cutting Quartz Glass

Paul Rathmill

Enterprise Q Fused Silica, UK

(1) Draw up profile required using CAD. This gives the required dimensions for the laser program to be written.
(2) Write laser program to cut the required profile and also for end plates to hold tubing in place whilst on the rotary axis. Program can be tested on plastic tube to make sure it is correct.
(3) Cut the quartz to the required length to fit on the rotary axis.
(4) Attach quartz tubing to the rotary axis and locate the laser nozzle in the correct position to run the program.
(5) Adjust the laser power and speed to the correct level to cut through the wall thickness of the tubing.
(6) Run the program. Laser should cut through quartz cleanly giving a polished finish to the cut edges. Speed and power can be adjusted to give the required finish.
(7) Once the program is complete, remove quartz from rotary axis and check dimensions are correct to the relevant drawing.
(8) Wash quartz in hydrofluoric acid to remove silica residue.
(9) Rinse in deionised water.
(10) Anneal quartz at the required temperature. Soak time can be adjusted depending on the thickness of the quartz.
(11) Check annealing process with stress gauge.
(12) Flame polish quartz after annealing. This removes sharp edges and micro cracks from the laser cutting process and will help prevent cracking during the next process.
(13) Repeat the annealing process.

Fig. 12.2.1. Quartz tubing secured to rotary axis.

Drawing on Experience 375

Fig. 12.2.2. Laser cutting slot profile.

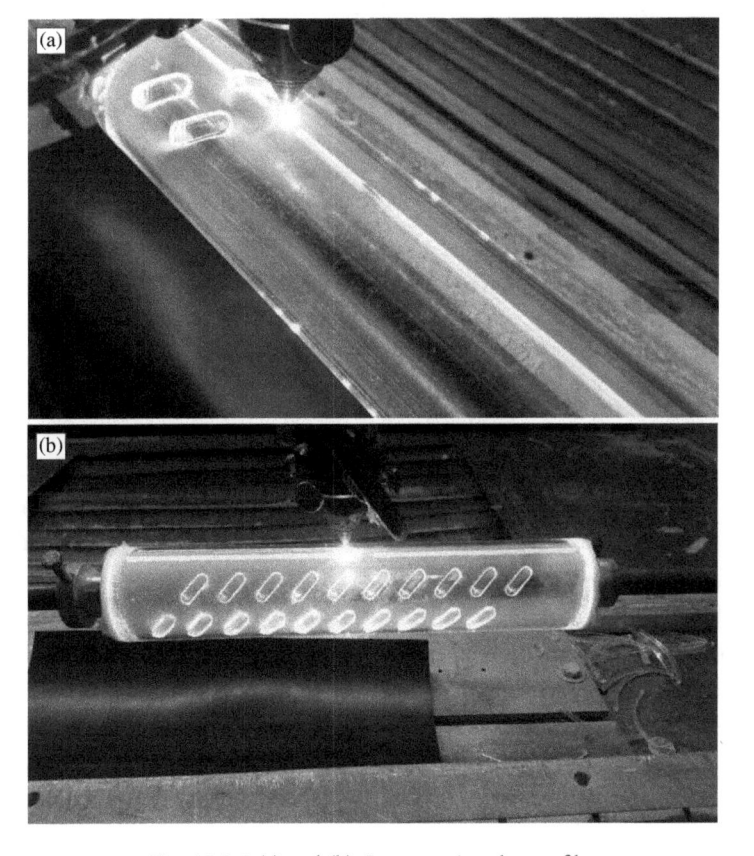

Fig. 12.2.3 (a) and (b). Laser cutting slot profile.

Fig. 12.2.4. Finished profile with polished slots.

12.3 Fit, Form and Function

William Fludgate, *FBSSG, Chairman*
British Society of Scientific Glassblowers, UK

Same items, many approaches, or is it Fit, Form and Function.

There are many times in the glassblowing workshop someone will ask for something that is not just a standard piece of laboratory glassware. This could be a simple request with a sketch on a scrap of paper to a full blown engineering drawing.

Let me explain why sometimes that simple sketch is more useful, it usually means that I can use a bit of poetic licence, knowing the material that we use, commonly it would be borosilicate glass.

I can then explain the types of stresses imposed in the manufacture that would come from its concept, to the fully made item and that it would need to be changed to achieve the end result. That means the person that requires the item is usually very happy with that and it make my life easier as well. Not that I am trying to get out of making the piece as they request, just trying to keep the costs incurred down and making sure the glassware survives the handling with the end user.

On the other hand, the engineers that produce some wonderful pieces of artwork using computers often forget that they are working with a material that most of them do not understand and do not take into consideration the extra hours of work that obviously drives up the cost, if it's possible at all to produce in the format they provide. This usually means that a meeting has to take place to amend such drawings, get that approved, issue another drawing number revision code and so it goes on. While that is happening, I would look at the drawing, try and make sense of what they hope to achieve and make something similar. They would come back with another wonderful computer generated artwork and I would hand them my cobbled up test piece. Often, and I really mean very often, they would "use" my glassware as a test piece and I would then get another drawing made to the specifications of the cobbled up version. That's why I prefer interaction with someone that asks for assistance and has a sketch plan.

On one such occasion, many years ago, it had been decided that a particular experiment needed the use of a Schlenk line, not just any Schlenk, this had to be a fully jacketed unit.

When I asked about the importance of the jacket, I was told that they are well aware of possible implosion or explosion whilst in use. This is true, if the procedures are not followed in its proper use.

The drawing was sent down to my workshop and I have to say, it was a feasible project on paper, but as a scientific glassblower, one often looks at glassware from the inside out, breaking down how one would start the making the components prior to assembly.

Now a standard line would consist of a dual manifold with multiple ports that have taps, what was given to me on paper was the equivalent of having a ship in a bottle, inside, a ship in a bottle that is what it looked like.

Now I have made many types of jacketed vessels over the years, from simple Liebig condensers, Allihn types and many types of coil condensers.

However, this was one that was causing me a problem. How to make the unit, and put it into the "jacket" before the taps were to be attached was not the issue, I had to hold the large tube, the jacket, in a large lathe, slide the unit inside the tube and warm the whole lot up, or it would crack. I resorted to wedging the pretapped Schlenk line into the jacket with copper strips and turning the large lathe on to turn very slowly. With burners lit beneath the large tube to warm the body up, I would stop the lathe and put the tip of two hand torches in each end to also warm up the inside piece, this was going to be a long process, while the burners were keeping the outside warm, I prepared the taps to be attached, oh did I not say there were eight taps and four ports to consider. The process of assembly started at 9.15 am, as each tap was sealed into place, and annealing the area of the work, the "warming" process had to be maintained. The last tap and port had been attached, the kiln had been prewarmed to 425°C, the copper strips were removed where I could reach them, carefully placed in the hot kiln and set the kiln to anneal the glass, I had spent over twelve hours on that part of the job.

It survived the kilning, the copper that I could not reach was removed easily with nitric acid. I had survived without any cuts or burns and I had to say that I was very pleased with my piece of work.

When the unit was installed, it worked as well as I would expect. I did not ask why the jacket was required in the first place, I just made what was requested.

About a month later I was called up to the laboratory to see if I could replace a large piece of equipment that had broken, no, it was not the piece I had described, that was on the side near the window. I asked were they not running that experiment anymore. The answer I received was astonishing, "Oh we used it once, we replaced it with the old one (a standard sized Schlenk line) we're now using that and keep it in the fume cupboard, that one looked too big." I walked away muttering to myself.

I had spent many hours scratching my head prior to manufacturing that nightmare, the preparation, many little drawings that I scribbled down, the worries of the "jacket" cracking and having to start all over again, only to be told; It's too big!

If you can take one thing from this article, it is this, the danger is not in the material we use, the hot flames, the types of acid that is in the chemical cupboard, and it's not that glass can cut you so easily. I deal with some of the most dangerous elements on this planet.... People!

12.4 Combining Artistic Glass Working Techniques within a Scientific Glassware Context

Ian Pearson, *FBSSG, past Chairman British Society of Scientific Glassblowers, UK*

Ian Pearson was trained as a scientific glassblowing and is a past Chairman of the British Society of Scientific Glassblowers. He has worked for various scientific glass companies before moving to Caithness in Scotland where he worked for the then UKAEA, now Dounreay Site Restoration Ltd as their senior scientific glassblower. In 1990 he started his own business in Thurso combining his scientific and artistic skills with the results that his sculptures have been featured in several exhibitions throughout Scotland. He teaches basic lamp-working skills at Northlands creative Glass at Lybster several times a year and has worked with several artists in residence there. Ian is currently Editor of the *Journal of the British Society of Scientific Glassblowers*, a role he has occupied for the previous 25 years.

Introduction

A lot of basic skills used in manipulating glass rod and tubing in a flame are easily transferred from the scientific environment to the artistic world and visa-versa. Evidence of this can be seen with glass bead makers and other functional art items where even turning of tubing is vital for quality results. Shaping glass using tools made of carbon is a common practice both with scientific glassblowers and in the artistic world. Indeed tooling necks of glass vessels with a carbon reamer uses an identical approach whether the item is a chemical flask of a perfume bottle. This article describes an approach for creating an item of scientific glassware but using techniques and a style of working that was developing through artistic flame working.

380 *Laboratory Scientific Glassblowing: A Practical Training Method*

The Technique

Photographs 12.4.1–12.4.3 show the technique which is the focus in this chapter, that is, heating tubing, positioning on a carbon "flat" and pressing with a hand held carbon paddle. A simple and quick method for shaping glass. Photograph 12.4.4 shows the author blowing down the end of a tube to create

Photograph 12.4.1.

Photograph 12.4.2.

Photograph 12.4.3.

Photograph 12.4.4.

a flat surface. To ensure square corners, glass rod is added to the desired position and whilst still molten the area is slightly overblown whilst at same time pressing glass with carbon. Photographs 12.4.4 and 12.4.5 show this technique being performed and the end result can be seen in Photograph 12.4.6.

Photograph 12.4.5.

Photograph 12.4.6.

The Thought Process

It is common practice to apply scientific thought to a scientific job. There are perceived differences in approaching science and art which may hinder an effective outcome. One obvious view is that perfect craftwork is paramount and nothing less than 100% correctness is tolerable. A popular approach to creating square or rectangular profile tubing is to manufacture a mould with exact dimensions of the required product. Machinery is then employed incorporating a lathe and perhaps using a vacuum system with cold working techniques such as cutting and grinding completing the whole operation. The technique described here for the specific item of glass uses no machinery or cold working tools. The end result was exactly what the customer was looking for and the item was made in a fraction of the time had other methods been considered. The thinking was generic and conceptual at the very instant of looking at what was required.

One aspect of working glass in a flame which both scientific glassblowers and artists share is that of stress. It maybe be said that scientific glassblowers are more aware of dealing with stress and the annealing process. One view is that scientific glass products are built for use whereas art is for viewing only. Anyone who has seen their favourite glass bowl or paperweight crack would surely disagree!

The late contemporary glass collector Dan Klein has written many books on art glass. A photograph in one title *Glass — A Contemporary Art* (ISBN 0004122283) shows a glass sculpture with a crack. On asking Dan why such an imperfect item of glass was included in the collection, the reply was that the item was still great art if not perfectly made. So the glass artists sees a crack in glass as a not a flaw where the scientific glassblower would disagree. The risk of commencing to work glass in a flame does differ between those with a scientific background to those who have their roots in art. Ian Pearson who has worked with many glass artists at Northlands Creative Glass Studios says that scientific glassblowers appear to be more precautionary in spending a lot of time preheating glass and warming vessels before starting whereas artists have a more straight forward approach and it is this latter style that was used in thinking how best to complete the item described in this chapter.

When handling complex shaped glassware whilst working on a section of that glass in the flame, one is very aware not to "brush" the flame along cold surfaces or allow the glass that isn't being heated to pass through the flame unplanned. This action could introduce stresses and create small cracks either on the surface or through the wall of the glass tube. Using a more artistic approach in thinking then one has the potential to gain confidence to be more bold in presenting all and any areas of the glass item to the flame with minimal of planning. In respect of the item of glassware described later in this chapter, then the work commenced in the centre and moved left and right with the minimal of preheating. Cracking did occur but because they were obvious, quickly seen and repaired as new joins completed.

Some glass artists working with flames favour the hand torch as a supplement to a bench burner. In the situation described here, only a bench burner was used. It is appreciated that for some people using a hand torch may have been easier but there is no guarantee that the end result would be superior. One main advantage of using a bench torch is enjoying the benefit of gravity but swinging the glassware through 360° and letting the glass flow in the desired direction. Jigs of course are available to enable glass to be clamped onto and turn through different angles but following the ideology of performance art sometimes it is the way the work is carried out rather than the end result that is important.

Ian Pearson has worked with two video artists who have used the act of shaping glass in the flame as the centre piece of a short video. It is the recordings of the work that is the objective rather than the work itself. In some instances, there is no tangible end result and only memories of the work exist.

The Item — Mixer Settler

Mixer settlers are a class of mineral process equipment used in the solvent extraction process. A mixer settler consists of a first stage that *mixes* the phases together followed by a quiescent *settling* stage that allows the phases to separate by gravity. At Dounreay, mixer settlers made of stainless steel were used to process irradiated fuel solutions. In an effort to study options of improvements with the process, glass models of various configurations were made.

Glass Mini-mixer Settler

The original design of a glass mini-mixer settler used standard tubing and rod with circular cross-section (see Photograph 12.4.7). A close up of the section where the test tube sections are joined can be seen in Photograph 12.4.8. The technique for this type of join is identical to that of squaring the corners of a tube with carbon as described earlier.

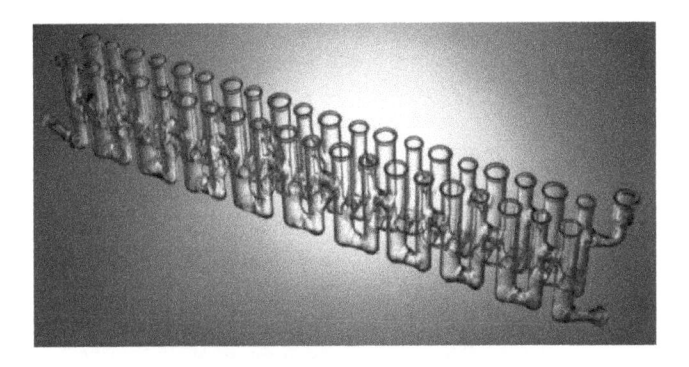

Photograph 12.4.7.

Photograph 12.4.8.

Glass mini-mixer settlers have proved invaluable at Dounreay for carrying out experiments on handling solvent extraction and have been used in many different environments including glove boxes as shown in Photographs 12.4.9 and 12.4.10.

One can see the vast amount of PTFE tubing that is fed into the glass and it was suggested that to avoid such material, glass tubing would be an improvement as the whole item would be more secure. Photograph 12.4.11 shows the author working on such an item which proved to be as complex to operate as it did to make. Thus this design did not prove practical.

However a need was identified to replicate the configuration of the actual stainless steel mixer settler used in the plant areas. This involved constructing a larger glass mixer settler and replacing tubing of circular cross-section with that of a rectangular shape. Consideration was given to either purchasing

Photograph 12.4.9.

Photograph 12.4.10.

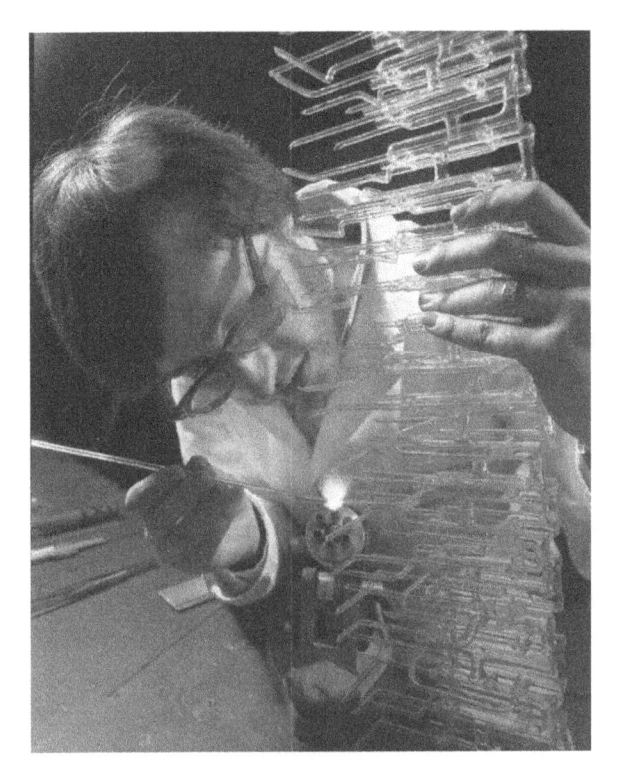

Photograph 12.4.11.

lengths of rectangular cross-section tubing from an external supplier or using moulds and a lathe to redraw circular sectioned tubing over a carbon mould engineered to specification. It was noted that the tolerance on the mixer settler was not onerous and tight specifications were not required and if employed may have allowed criticisms of over engineered to be made. A relative cheap and simple method for making short lengths of tubing with a rectangular cross-section was needed.

Construction of Short Lengths of Tubing with a Rectangular Cross-Section

Using 38 mm medium wall tubing, points were pulled about 300 mm long. The point was then heated in the flame and the glass pushed together whilst blowing slightly. This affected the glass tube to increase in diameter and the wall thickness to increase. This technique was repeated several times before the point was placed on a section of flat carbon and the operator brought down a carbon paddle on top of the hot glass whilst blowing at the same time. In this particular situation the size of the cross-section was to be 60 mm × 25 mm and therefore the operator knew to judge the distance between the carbon paddle and the flat carbon to be 25 mm. It is important to move the carbon paddle and the glass around in a circular fashion to minimise mould marks or lines.

Once a rectangular shape of glass was completed one of the spindles was removed in the flame and the end of the tube flattened. As normal practice focuses on circular tubing, with even turning, shaping rectangular tubing would not benefit from this approach, so a more "stop/start" method was used. It is important to ensure that the corners are sharp to maintain the appearance of a rectangular and not allow gravity to dictate the circular result which tended to transform. To emphasize the sharpness of corners, glass rod was added to each corner in the flame and the glass blown slightly before surplus rod removed in the flame.

Side-arms and additional supports were joined onto the side of the vessel before the top was pulled off and any unevenness flattened in the flame.

Photographs 12.4.12 and 12.4.13 show the completed item. The size of the item can be appreciated by the inclusion of a hand in the view.

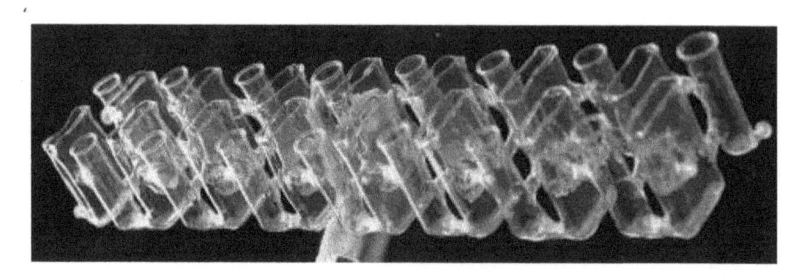

Photograph 12.4.12.

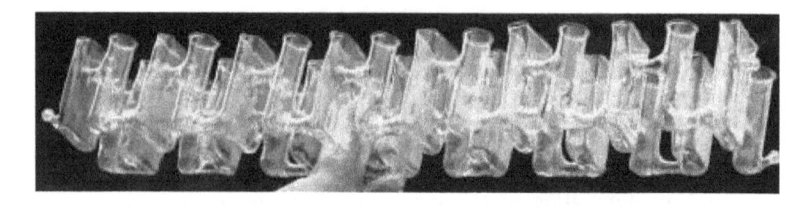

Photograph 12.4.13.

Conclusion

The completed item satisfied all design criteria but at same time recognising flawed appearances. In approaching the work from a non-scientific manner, the character of the glassblower was tested and confidence was gained through increased courage to tackle more complex shapes.

12.5 The Foster Cell

Gary Coyne

California State University, USA

As a glassblower for over 40 years and a university glassblower for over 30 years, I have been involved with designing a number of items. The typical process is a professor comes to me with an idea, perhaps some photocopies or a hand drawing of an apparatus that he or she has found that's close to what they want and needs to be made with or without alterations. At that point, we discuss the good and bad points of the design and how to make it more robust (aka more survivable with students) and anything else I can come up with to improve the final product.

In 2002, a professor came to my office with a collection of criteria but no thoughts on how to design the apparatus. The requirements were that she needed to pass a gas through a chamber that was water-cooled and also needed to shine UV light on the condensate to cause a reaction. We discussed the size and volume she was expecting to need and other dynamics that would help narrow down the design.

The UV transmission requirement meant that part (or all) of this was to be made out of quartz. Unfortunately my shop was not really set up to do much quartz work and if I could come up with a way to avoid a complete quartz cell, all the better.

I mulled on this some and came up with a solution that turned out to be relatively easy to make and did exactly what it needed to do. As a bonus I was going to make two cells at one time by making two parallel cells and cut them in half. Than each half would be its own cell.

I had come up with several different approaches to make this and after mentally going over the entire construction process, boiled it down to two construction ideas but then quickly settled on the second one. The top one in the figure below was one of the finalists but was quickly rejected. The idea was to

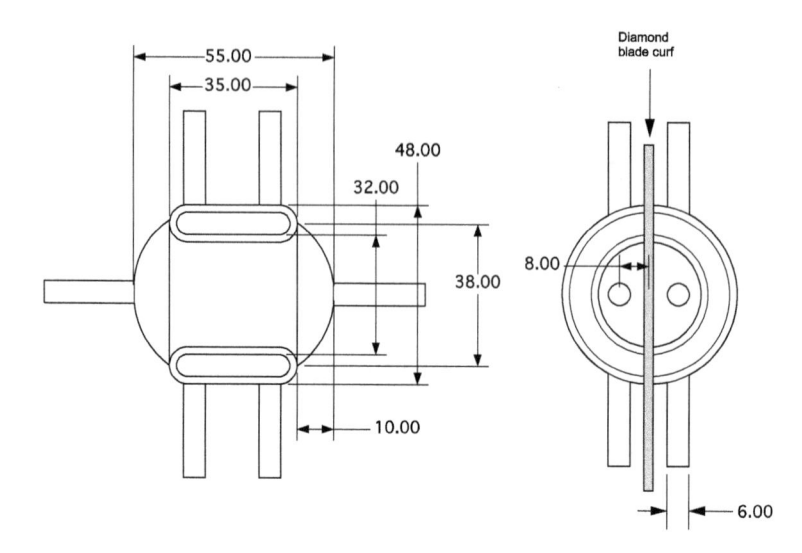

390 *Laboratory Scientific Glassblowing: A Practical Training Method*

make a thimble-shaped piece glass and support it on a smaller tube with carbon tape. This would then be sealed within the outside of the cell but I quickly realized that blowing into the various parts of the cell would not be possible.

My second approach was to start off with the two end tubes sealed onto a support tube held in the headstock of my lathe. This would be sealed onto the inside piece of the cell. Then, a Maria would be formed to seal onto the outside section of the cell, and the very end would be flared open. The far end of the cell would be sealed later once all of the sidepieces were sealed onto the body.

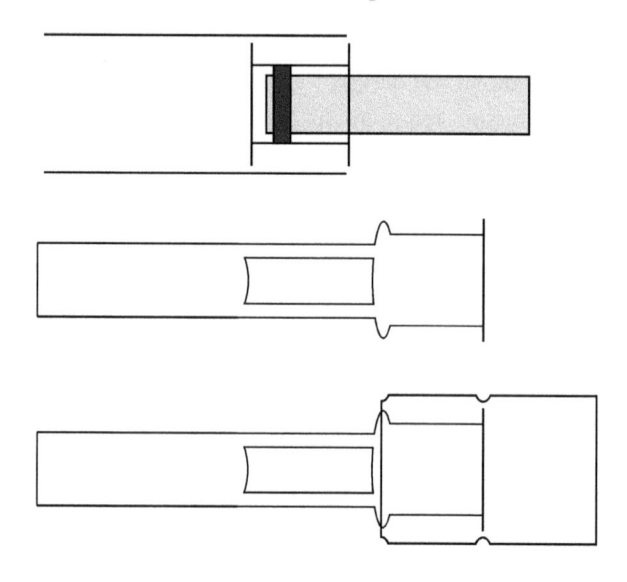

Now that the construction plan was determined, the assembly process was as follows:

Because the outside distance of the two 6 mm tubes was 22 mm. I used a 22 mm tube to first attached the two centre tubes and take advantage of aligning them with the outside of the that support tube. Then I could bring in the 32 mm tube from the tailstock and seal those two tubes onto what would become the inside section of the cell.

Now I could form the Maria for the ring seal, flame cut off the end, and flair it to size.

Also note the two extra tubes sealed onto the support tube shown below, these will be used to bring air into the centre region during construction as well as provide support for the various arms from the heat of the flame annealing. I also switched from tubing to rod to add these additions because it took less time.

Now the 38 mm tubing was placed in the tailstock and prepared to seal the ring seal. The far end of the inner tube would float as the seal was made and remain open until later in the construction.

Since I could access the air in the inner section because the right side had not been closed off, I started sealing on the side tubes and connecting them to the outside air tubes discussed in the previous step. As stated, these side tubes also provided support for my air-gas flame being used to flame anneal as the work progressed. This double-duty was a big asset. I only needed to have one of these tubes officially connected to the side tubes and simply used rod to provide the subsequent support during the flame annealing. As such,

I didn't need to concern myself with properly made seals on these arms and only needed to make connections to support the glass.

Once I had all eight side-tubes sealed on, I could then start to bring down the 48 mm tubing down to the 32 mm inner tube on the right (tail stock) side to complete the cell.

Once the end was shaped, I then added the last two end tubes, again adding support rods to keep things in place during the constant flame annealing.

The completed cell prior to formal annealing.

My next task was to cut this down the middle and that proved to be a new challenge and also provided some excellent new opportunities.

Drawing on Experience **393**

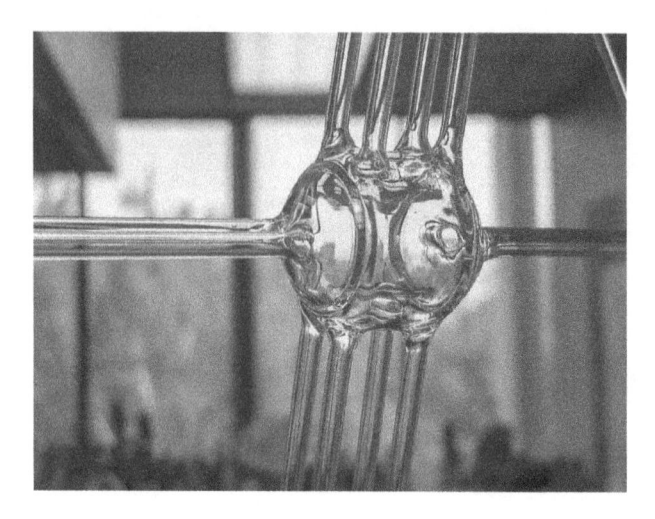

Source: Originally appeared in *Fusion* Vol. 53(4) and is republished with permission.

Like most glassblowers I have known, I had been making simple wooden supports for use in my wet saw. The problem is that wood warps and considering the potential fragility of this cell, I didn't want to have it destroyed because my fixture warped.

I have to add here that one of my hobbies is woodworking and in several of the magazines I've received over the years talked about a commercial, man-made product called Corian®. Corian® is typically used for countertops and other decorative hard surfaces and was advertised that it "cut just like wood." I always found that a strange advertising slogan to put in a woodworking magazine, after all, was not I trying to learn about working wood not some plastic?

But this had me thinking: could Corian® be used as a support for items in the wet saw? I contacted the company and asked them for some samples. I told them that I would be happy with anything they sent me, as the specific look of the item was not critical for my intended use. Rather than looks, I need to test its properties. They didn't let me down, one of the samples they sent me I affectionately called "Clown in a blender."

However, the original claims were completely justified: I could cut it easily in my table saw and shape it easily on my router table and drilling holes as well was very easy. Corian® is a brittle material (do not drop it) that cuts very well with either high-speed steel (good) or tungsten carbide (better) tools.

I did a variety of experiments to see if there would be any swelling or warping and after several days of soaking all or parts of the material in water,

Corian® proved to be 100% stable in water. It has proved to be the best material I've found to date for jigs and fixtures for the wet saw.

The cell-cutting jig, shown below, shows a channel for the cell to sit in with four holes to receive the water tubes. The two outside holes in the fixture itself were originally added as I was assuming that it would be best if I bolted this to the sliding table on my wet saw. As it turned out, once the flat surface of the fixture was placed on the wet surface of the sliding table, it didn't move and I never used any bolts to hold it in place.

Source: Originally appeared in *Fusion* Vol. 53(4) and is republished with permission.

When I originally planned this all out, I created a set of aluminum jigs to be placed against the cell during construction to help align and place the various side tubes. This turned out to be completely unworkable as it meant that I needed to hold the torch in one hand, the tube in the other hand, and my feet just were not up to the task to hold the aluminum jigs for their task. I ended up having to just wing it by eyesight.

This also meant that the holes I had originally drilled into this fixture had to be increased in size to get all four side-tubes in place. Because of concern of movement during construction, I used a goopy soft glue to hold things in place for the cut. When I made a second pair of these cells, I didn't bother with the goopy glue and didn't suffer any consequences.

As far as any accuracy in where the side tubes were located, that did not make any significant difference, as my main goal was to cut the cell in half.

A millimetre or two one-way or the other was not a significant concern. As it was, the result was surprisingly down the middle.

After cutting, I removed most of the goopy glue with a razor blade and burnt the rest off in the annealing oven overnight.

Source: Originally appeared in *Fusion* Vol. 53(4) and is republished with permission.

At that point I used an epoxy that was designed for use in a high UV environment to seal the quartz plate over the entire cell.

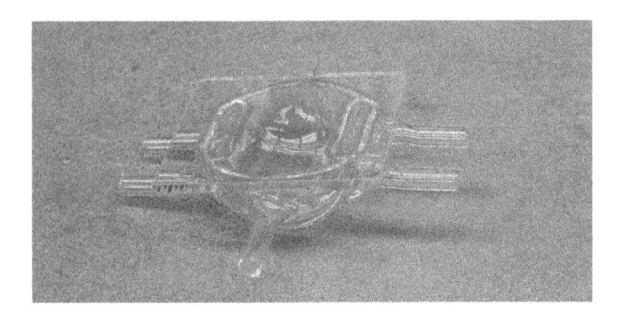

Source: Originally appeared in *Fusion* Vol. 53(4) and is republished with permission.

An Epilogue

At this point I do want to digress back to use of the Corian® material as a jig and/or fixture for the wet saw. While my work on the cell was done over 10 years ago, I continue to use the various tools I have made from Corian® ever since.

One thing to point out is that since it is a manmade material, it's absolutely consistent in its dimensions. The importance and value of that is that rather than laying down one 12 inch long piece on my saw to support a 12 inch piece of glass, I can lay down two or three pieces and lay my glass on that. The material itself is very hard and is easily cut by a diamond or carborundum blade.

Probably my "most used fixture" that I've made over the years has been a drip-tip cutting fixture as shown below. It was simply made by cutting out a small square and then cutting that in half across the diagonal (accuracy for this is eyeball only). Then using a simple epoxy glue, place these two triangles against the corner of two pieces of Corian® and let harden. Then cut into this on your wet saw just-off-of-centre.

This fixture can be used for both individual tubing pieces as shown below.

Source: Originally appeared in *Fusion* Vol. 53(4) and is republished with permission.

Or, after cutting a notch in the back piece to make room for the stopcock's barrel, the drip tip is directly on the stopcock. This provides a beautiful diamond cut end on my drip tips.

Source: Originally appeared in *Fusion* Vol. 53(4) and is republished with permission.

The point here is that by being challenged to provide the design of an apparatus that had numerous challenges, I also was motivated to discover a new use for a material I was already aware of for some time. Not only was this material superb for the original use but it also exceeded my expectations for a whole variety of alternate uses all of which were uses that Corian® was not originally marketed for.

Epilogue: The cell was successful as designed, it provided the professor the constant temperature she needed and the irradiation process did provide the results they expected and needed. Unfortunately, her work moved into other areas so that this line of research was never pursued further. In other words, that's science.

12.6 The Design and Construction of his Masterpiece

Konstantin Kraft-Poggensee, *(Master) past Chairman*
German Scientific Glassblowing Society, Germany

Introduction

This section is about my masterpiece I designed and built as part of my Master's exam in 2002 — a long time ago. I chose to write about this piece, because it taught me very much — about skills and life.

However, since so much time has passed since the actual practical work, I am not able to provide so many technical details about how the piece was made. It would be a bit too complex anyhow. So be warned that this is more a reflection on the piece, the problems that came along and what it meant to me.

The Young Master's Grey Hair

Glassblowers know what they need to do to become Masters — you need to work hard, learn from failure and persist. In Germany, you can actually take an official course and an exam from the Crafts Chamber. To be called a Master of Crafts, you need to pass the exam. And you need to become a master to be allowed to run a scientific glass-workshop and have a trainee.

I followed my Master's training in 2002. Part of the final exam is a masterpiece — a glass apparatus you sketch, submit to a committee for approval and then need to build within three months. The finished piece is then presented to the same committee with an ink drawing, a description of its function and a calculation of its market price.

When I had to submit the sketch for the masterpiece, I was offered to choose from a bunch of old drafts that were complex enough to be approved by the committee. At that time however, I was very disappointed with the whole master-thing. From that master course, which had cost me a lot of money in fees and three months of unpaid holiday, I had learned almost nothing. And these drafts were as lame as the rest — so I refused.

Now this was the only part of the exam I had an influence on, and the only point when I could do something differently and not just what was necessary to obtain that lifeless certificate. Something that would make it worth it in my own eyes to call myself a master glassblower. So I designed my own piece — a very big and complex distillation-head. It looked very impressive on paper — but it was way over my skills. It was a nightmare to build — and I had to change the design from the original draft. I managed to present it in one piece and it was outstanding in comparison with the pieces of my fellow master-students. The committee was impressed but had to follow the rules. The differences between the draft and some untidy spots led to a low grade.

What an irony — but two important lessons learned! *On the one hand, when it comes to an exam — it is probably not wise to make your life so much more difficult than necessary.*

And on the other hand — push yourself with challenging goals and be proud of your skills and not of your grades!

The Design Idea

I can easily say that the designing phase was the most fun part of this whole adventure. For sure, that was part of the problem on this piece — I got carried away and added more and more ideas into it. I spent hours and hours to figure out dimensions of tubes and the steps in which the piece could be done. It was a delightful puzzle, using the tubing catalog as a tool. Because so many elements were so tight together — I had to make sure tubes of the right diameter and wall thickness would be on the market.

It was supposed to be a distillation head — but a special one: its function evolved with the creative process. One of its parts, the integrated upper-column with sample and measure points, was meant as a small copy of the whole distillation unit.

The idea was to use the results of the sample and measurement points to adjust the distillation process. Parameters like the temperature in the boiling flask, the flow of the cooling water and the back-flow of the distillate would be adjusted automatically. For example, there is an integrated moving funnel that can be triggered with magnetism. This can let the distillate either back to the main column or into the after-cooler.

Whether this design idea made any sense and whether it would have worked, I have no idea. The main purpose was to create a great masterpiece. And for sure this worked.

Some of the Features

Yes — I really got carried away with the design. But this is the part I'm really proud of. I managed to put a lot of things into the piece that I really wanted to build one day. Here are some of them:

A double-walled flange: This was one of the last things to be added to the piece and it did not really work out. Especially since I had to grind it attached to the almost-finished piece.

A fractionating column with three bubble-cap trays and a vacuum jacket: This is the piece with the sample points. I really think it is fun to build these; it is complex and lots of work, but it is not difficult.

A moving funnel: I really like the idea that you can switch a glass part inside a glass body. You need a funnel that swings on a glass rod. On the funnel, there is a small steel rod sealed in an attached glass tube. An electric magnet outside the glass body can then pull the funnel towards it. Again, it is a lot of work, but it is not very difficult.

A water-cooled stop-cock: There are three jacket stop-cocks in the system, two one-way together with the moving funnel unit. There is also a two-way self-designed stopcock condenser. To build this, I really started from scratch, both to design it and to build it. But it turned out great and it ended up being the piece I like most about the whole apparatus. It was difficult and required many steps and the use of different techniques.

A coil condenser with many coils and a double jacket: This was a bit of a puzzle but it was not so difficult either. And I'm proud of the design with the cooled dripping spear that also connected the coil pack with the jacket.

When I look at this list, I wonder what was so difficult. But then it was to put it all together, with the constant fear that the stress in the glass would be too much or to hit a cold spot with the torch. And all the little problems you never thought of beforehand.

The Coil Incident

Here is one example of the struggle with tight dimensions. The piece was designed to fit into a 100 mm OD, 3.5 mm wall tube. This also included the coil condenser part, which contains a pack of three coils with a tube inside. I had carefully chosen the diameters of the tubes and the inner diameters of the coils, and I knew it would be a tight fit.

However, there is always a slight difference between what you calculate and what you fabricate with your hands. For example, the finished pack of coils didn't fit into the 100 mm tube where it was meant to fit. A bigger tube was not available and would have ruined my straight design.

The solution might not have been very elegant, but it worked — I used a belt grinding machine!

With a fine belt, I ground off a bit of the outer coil to reduce it to a smaller diameter. That was not a lot of fun — since the pack of three coils wobbled a lot and did not feel very stable. But I took it slow and made it fit. To hide the ground surface, I polished it with a cork belt — which pulls hard on the glass and made it even less fun to do. But I got my pack into the tube — and once it was in there, you couldn't have guessed that all these little corrections had been necessary.

Unclean

When you make such a piece once in your lifetime, you want the piece to look great — and clean: one should have a clear vision of this whole complex interior. However, I had never anticipated that cleanliness could be an issue.

Then suddenly it was very clear that this was an issue: Since the piece had so many seals and passageways, I constantly got condensate from the torch into it. Once it dries out, it creates some white traces — everybody knows this. Every time I got the piece out of the annealing oven, there were more of them.

I was on a tight schedule — rinsing a piece like this and drying it out (for sure you do not want more water than necessary when you work with the flame on the piece — often enough some condensate reaches the wrong spot and destroys something beyond repair) would have thrown me back an extra day. My idea was to rinse the piece with citric acid when it was completed.

When the piece was finally completed, it looked very precious and really big; I was even scared to carry it around. There was no way I could rinse this thing — handle the whole piece while wet. Maybe shake it with acid or water inside? Drain the water into a sink? How could I possibly get all the water out again without causing more stains? My nerves told me to swallow the pill and to present a dirty, but intact masterpiece.

The Drawing

One thing I did not foresee when I designed and sketched the piece was the complexity of the technical drawing. Building this one piece from a sketch was a challenge — but part of the presentation for the exam was a technical drawing that would allow a glassblower to build the same exact piece without asking any further question. Every measure of every little detail needed to be on one page — neatly drawn in black ink. It seemed impossible to arrange all the dimension lines. I figured it out — with very many detailed drawings. If I had envisioned how difficult this one task would be, I may not have been temped to choose this design as my masterpiece.

Letting Go

A bit more than 10 years after I made my masterpiece, I decided to quit my job and move to Norway. Every cubic-metre of belongings moved cost me dearly, so it was a good time to clean up and almost everything that was not useful had to stay behind.

Now there was a riddle. My masterpiece never made it out of its big box after the presentation to the jury and was catching dust. It was never used and would never be. Still — it was the biggest, most complicated and important piece of glassware I ever made, or will ever make.

I am not a trophy guy. It was in a box all that time — and there was no reason that it would be any different in another country. So against everybody's advice, it went into the trash. And that felt really, really good!

Conclusion

When I designed the piece, I obviously miscalculated my skills. By writing this text, I realized one underlying cause. When I designed the piece on paper, I developed the way to build it as I went along. Again and again, at every step, I went through the process of how I would actually build each piece to see if it was possible. Only when I thought it was possible, I added it to the design. However, I never pictured what it meant to build the whole piece. If you work on a passageway on a virgin tube, it is one thing. On a big piece, with hundreds of hours of work in it already, where there are many spots around, you should not hit with the flame or get too hot again, it is a different story.

I am not sure whether building this piece made a real master out of me. There are so many glassblowers out there who are more talented than I am. But it had the effect on me I intended it to have: I really learned a lot and it was the part of the master class that made me feel worthy of being called a master.

A funny part of the story is that the piece was done again by a colleague as his masterpiece. He saw it an article I wrote for the German glassblowing journal. He was much more experienced with big complex apparatus than I was as a Master student, and wanted to build something challenging. He asked me if he could copy it; I said yes and sent him the drawing. He built it absolutely perfectly — without needing to ask me any questions. At least the drawing must have been pretty good.

I think I can conclude that I did everything right — including throwing it away in the end. Because I never missed it. And it is not important anyway. The important thing is to know what you are capable of.

Fig. 12.6.1. Rebel master with his piece.

Drawing on Experience **405**

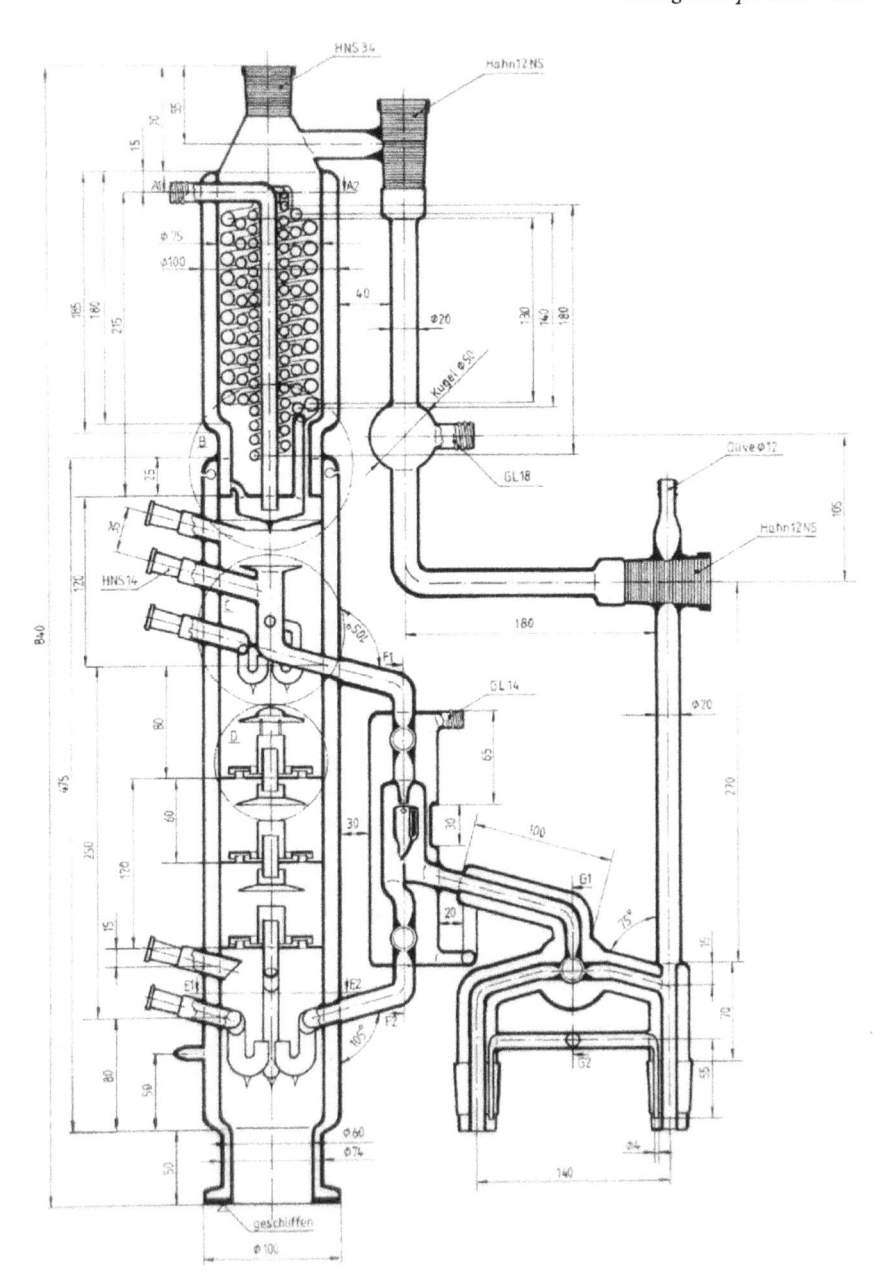

Fig. 12.6.2. Drawing madness — it took many detailed drawings to capture the whole piece.

406 *Laboratory Scientific Glassblowing: A Practical Training Method*

Fig. 12.6.3. The main drawing of the whole piece.

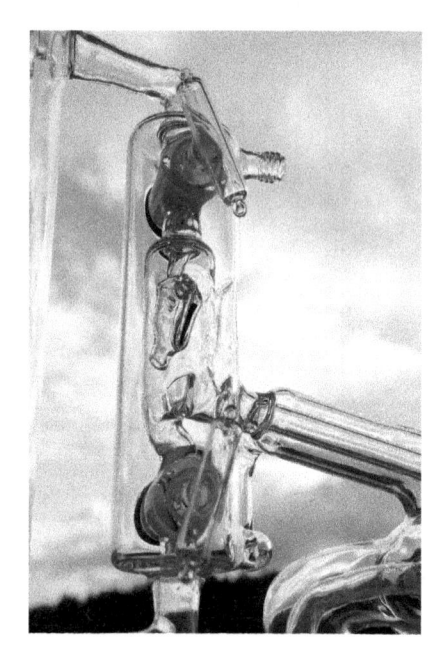

Fig. 12.6.4. The piece with the moving funnel.

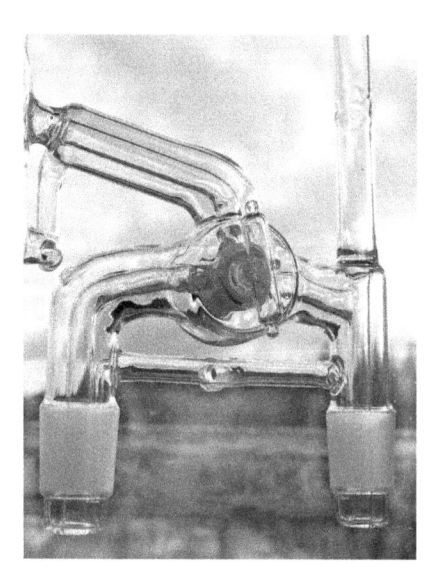

Fig. 12.6.5. The two way jacket stopcock condenser.

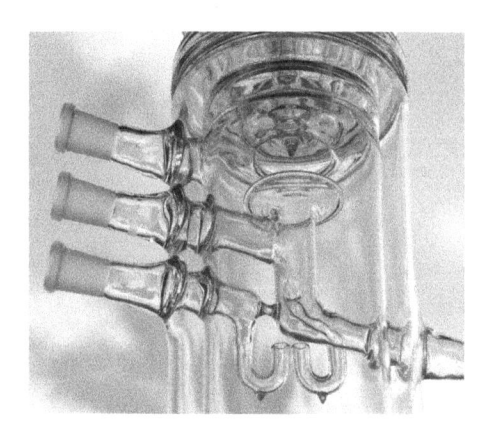

Fig. 12.6.6. The finished piece.

Fig. 12.6.7. The end of it all.

Fig. 12.6.8.

12.7 Variable Temperature Gas Inlet for Surface Science Ultra-High Vacuum Molecular Beam System

Keith Holden, *FBSSG, President*
New Zealand Scientific Glassblowing Society, New Zealand

This work was undertaken in 1977 for the Surface Science Group, Department of Inorganic, Physical and Industrial Chemistry, University of Liverpool, headed by Prof. D. A. King now Prof. Sir David King.

Introduction

A. Molecular Beam Techniques: A method for studying gas — surface collision dynamics
B. Gas Inlet System for borosilicate glass molecular beam apparatus

Design Considerations

(1) Fabrication Material — borosilicate glass COE 33
(2) Operational Temperature Range, 77K–723K (–190°C–450°C)
(3) Optical windows (2) for laser beam alignment of gas stream directional capillaries (2)
(4) Glass to metal seals (3) centreless ground tungsten rods. High voltage tesla coil discharge conductor rods for capillary jets (2) possible obstruction clearance
(5) Accommodation of temperature range fluctuations by way of glass bellows (2) and glass coils (2)
(6) Graded seals, borosilicate glass COE 33 to fused silica (2)
(7) Demountable nitrogen gas flow variable temperature control unit

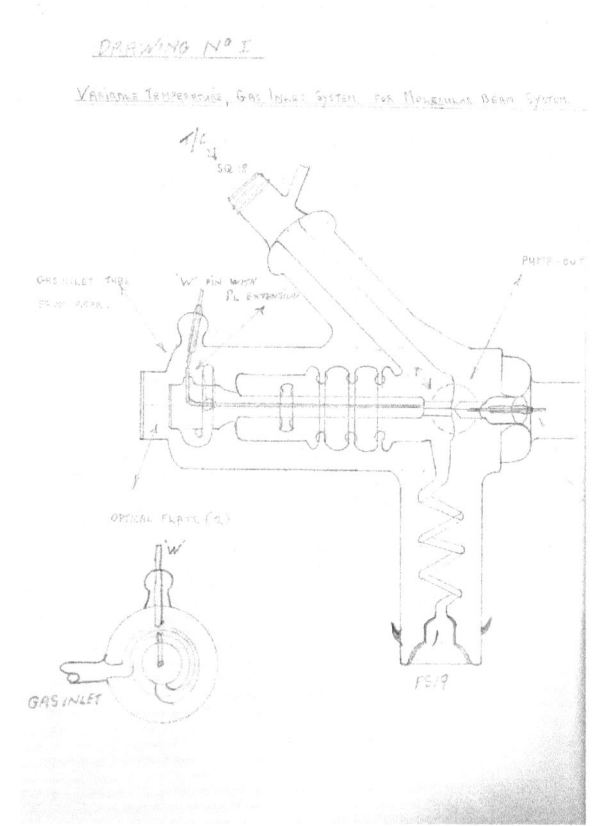

Drawing 12.7.1(a).

(8) Thermocouple temperature measurement at inlet capillary
(9) Precision bore capillary (2) 0.1 mm and 0.5 mm gas inlets alignment by laser beam

Design

Incorporating the requirements as outlined, numbers 1–3 and 5–9. Drawings 12.7.1–12.7.8 show the following:

A. Internal coils (2) for coolant/heating nitrogen gas flow inlet via spherical cup joint size 19 and vacuum system sample gas inlet connection
B. Expansion/contraction bellows (2)

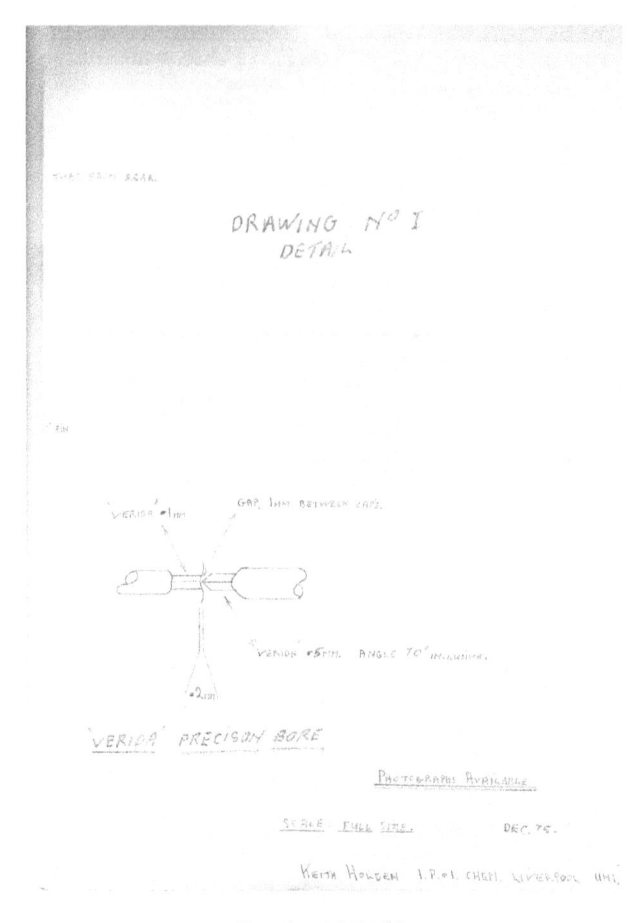

Drawing 12.7.1(b).

C. Optical windows (2) external and internal, for laser beam capillary alignment
D. High voltage tesla coil discharge pins (3)
E. Variable temperature nitrogen gas flow demountable delivery unit (see Photographs 12.7.1 and 12.7.2)
F. Improved design Mk II Unit

Fabrication Method

1 *Capillary 0.1 mm/single internal bellow/optical window section*: The 0.1 mm capillary is blown out at one end to form a bubble, which in turn is blown open and joined to a length of 8 mm OD light wall tubing. The

Photograph 12.7.1.

capillary is cut to a length of 13 mm. This capillary end is again blown into a bubble and flared to form a flat flange.
2 *Triple Bellows*: 24 mm OD light wall tubing is formed into a triple bellows section leaving an open end 20 mm from the bellows. The single bellows section is placed within the triple bellows tube, packed in place by using a constructed glass tube holder.

The 25 mm diameter outer tube is tooled down onto the internal capillary tube flange and fused together. A 10 mm diameter tube is sealed to the created dome end at a 45° angle. Opposite this tube at 180° seal on a 7 mm OD tube.

Optical Window and Tungsten Pin

With the packing support tube removed the assembly is placed in the lathe, blowing attachment connected to one of the two side arms previously joined in place.

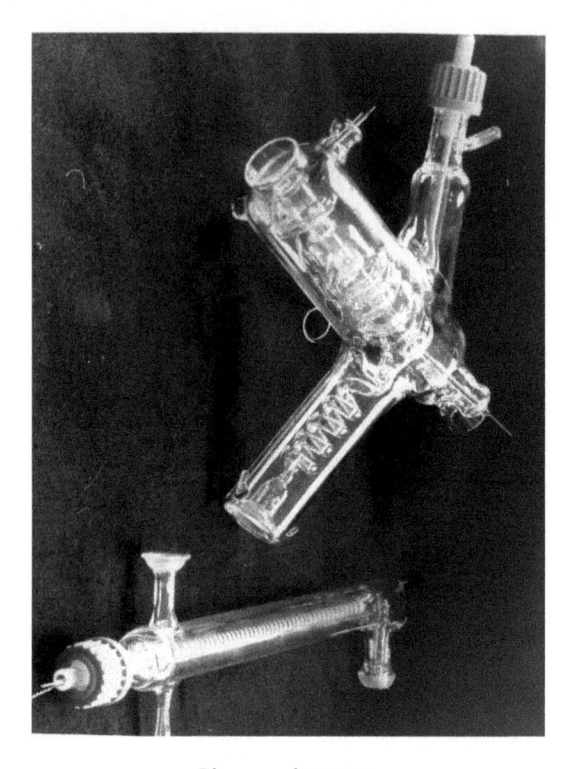

Photograph 12.7.2.

The 24 mm OD tube is heated and tooled down to fuse to the inner bellows tube, 16 mm diameter tube joined to the seal. A preformed single turn coil of 5 mm diameter is joined to 16 mm diameter tube. This serves as the entry port for the experimental gas in flow.

Fuse in place a stub of 10 mm diameter tube, 6 mm long. Insert into the stub a tungsten pre glass sheathed 2 mm diameter pin with a platinum wire extension. The platinum wire is bent to a horizontal position terminating in close proximity to the 0.1 mm capillary. Fuse the tungsten pin in place at the 10 mm diameter stud end.

The 16 mm diameter tube is flame parted at a distance of 5 mm from the tungsten pin.

A pre-prepared 16 mm diameter optical disc, chamfered edge, is held in place a carbon rod vacuum device. The window is sealed in place. Flame anneal, remove from lathe and furnace anneal.

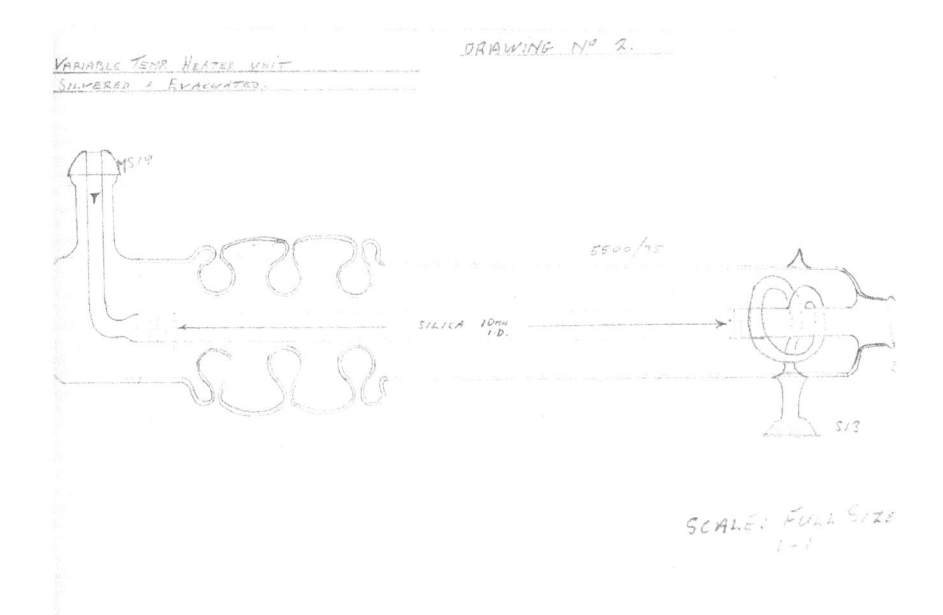

Drawing 12.7.2.

Drawing 12.7.3.

DRAWING Nº 5.

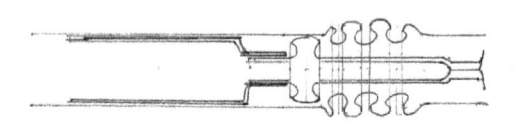

SCALE: FULL SIZE

Drawing 12.7.4.

DRAWING Nº 6.

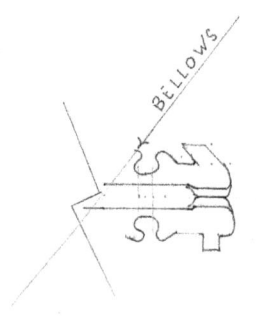

BELLOWS

SCALE: FULL SIZE

Drawing 12.7.5.

DRAWING Nº 7.
INNER ASSEMBLY

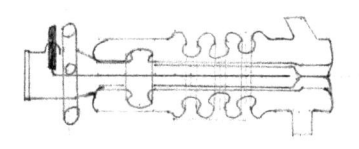

SCALE: FULL SIZE

Drawing 12.7.6.

DRAWING Nº 8.

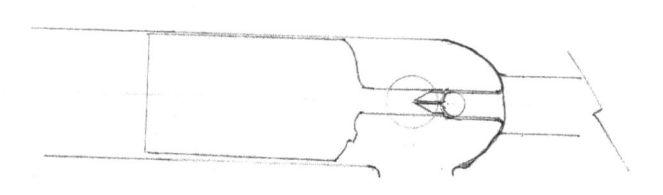

SCALE: FULL SIZE.

Drawing 12.7.7.

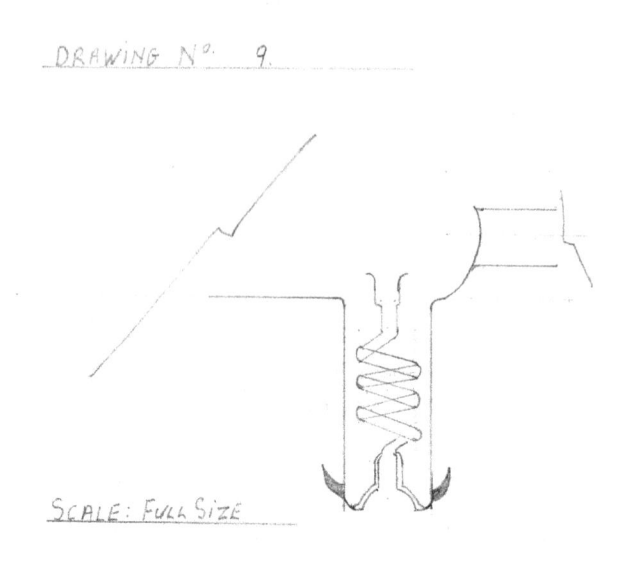

Drawing 12.7.8.

Outer Body Preparation, Assembly and Completion, Component Parts Preparation

A. 25 mm optical disc, chamfered edge
B. 2 off tungsten rod, 2 mm diameter × 30 mm electrode pin, glass sheathed with appropriate tungsten sealing glass. A button maria sealed at one end of glass sheath
C. Form a 5 mm diameter light wall tube coil, 3 complete turns on a 10 mm diameter mandrel or free hand form. Coil tube end re-aligned to be horizontal and central to the coil. Join coil to a size 19 spherical joint cup. Total length 70 mm
D. 30 mm diameter light wall tube, closed one end, length 100 mm
E. 24 mm diameter tube, 100 mm long, closed on end, opposite open end cut or grind to a 45° angle. Fine grind tube end, wash and flame polish
F. 2 off 10 mm diameter tubes, closed one end
G. Outer jacket tube 44 mm OD light wall, one end closed. Length 200 mm
H. Precision bore capillary tube 0.5 mm join to a 10 mm diameter tube. Cut capillary to length of 10 mm. Grind a 70° taper to cut end. Cut to overall length of 30 mm

Assembly

0.5 mm Capillary

Select glass tube that the capillary fits into closely. Join this tube to another glass tube that fits inside the outer 44 mm OD tube, use this as a holder tube, place inside the 44 mm diameter tube, seal the capillary tube to the dome end of the 44 mm diameter tube, attach a 20 mm tube.

Join a 10 mm diameter tube at an angle directed to the taper tip of the capillary. Later, this tube will have a tungsten pin sealed in place, the tip of the pin directed in close proximity to the capillary taper tip.

Join on a 18 mm diameter tube to the 44 mm diameter outer body 90° opposite the capillary. This tube serves as the UHV pumping connection.

Create a hole 30 mm diameter at 90° to the pumping tube, flame anneal. Remove the internal capillary holder tube.

Insert the completed inner assembly into the outer body, back in place, copper gauze can be used to be later removed by dissolving the copper in nitric acid. This operations is to be undertaken in a fume hood for harmful vapour extraction.

Align the 0.1 mm capillary with the 0.5 mm capillary, also space apart by 1.0 mm. Alignment checked and confirmed with aid of a laser beam directed horizontally along the tube axis. Spacing can be confirmed by use of engineers feeler gauge, 1.0mm thick. Access through the outer body 24 mm diameter hole.

Ensure the inner assembly 7 mm diameter tube is located in position central to the outer body pre-prepared 24 mm diameter hole. When alignment of capillaries is confirmed, complete sealing operations as follows. In the 7 mm diameter tube stub located central to the hole, join on the coiled size 19 spherical cup joint section. Place over the cup joint the 30 mm diameter tube. Seal in place to the outer body hold. Proceed to form a 24 mm diameter hole opposite at 180°. This will expose the internal 10 mm diameter thermocouple access tube. Extend this tube to a length of 60 mm. Place the 24 mm diameter 45° angle cut tube over the 10 mm diameter tube and seal in place. Flame anneal *in situ*.

Close the 24 mm diameter tube over the top of the internal 10 mm diameter tube, seal together and seal on a size 18 screw thread tube with the addition of a gas flow outlet tube, 8 mm screwthread. This section becomes the thermocouple entrance arm and coolant gas outlet. Flame anneal. Size 19 spherical joint.

DRAWING N° 10.

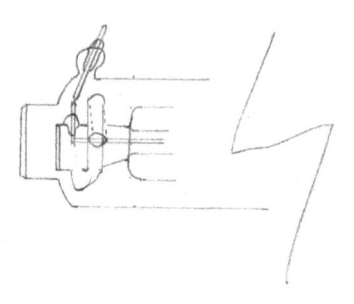

Drawing 12.7.9.

Cut and remove the 30 mm outer tube opposite the spherical joint rim. Seal tube to the size 19 cup rim. Ensure not to distort the ground joint surface.

External Optical Window

Tool down the outer body tube to 25 mm at a distance of 25 mm from the internal optical window. Flame part the 25 mm diameter tube and proceed to seal in place the 25 mm diameter optical window. Locate and hold the window in place by use of a carbon rod drilled out and connect to a vacuum pump.

Tungsten Electrode Pin

Opposite the internal tungsten pin, join in place a 10 mm diameter tube stub, seal in tungsten pin and locate pin tip in close proximity to the internal pin. These two adjoining pins serve as the tesla coil discharge application point.

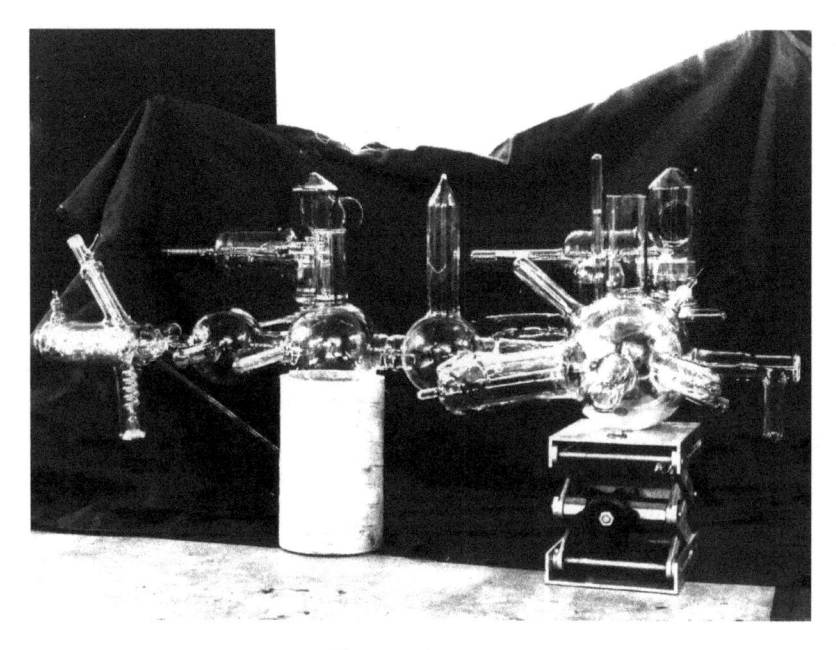

Photograph 12.7.3.

Experimental Gas Inlet Arm

The internal single coil gas inlet tube, fuse to outer body and join on an extension tube (see Drawing 12.7.9).

Completion

The completed assembly is furnace annealed, attached to a vacuum system and tested for vacuum integrity.

Installation

The completed unit placed *in situ* horizontal and aligned to the molecular beam apparatus. Capillary, 5 off of increasing diameter and target crystal confirmed as in alignment. The assembly is sealed in place. The completed and assembled molecular beam system is located within a retractable oven and further connected to three high performance mercury diffusion pumps. Pumping out and vacuum baking at 450°C achieving vacuum 10–10 torr or better (see Photograph 12.7.3).

12.8 Former Apprentice of Seven Years

Phil Murray, *Churchill Fellow, apprenticed to the Author for 7 years*

Where I Started

Lessons by their very nature have a predetermined outcome. School science 'Experiments' are far from experimental — hence I guess why they began to be called 'practical sessions'. Throughout high-school/college I had proved that I had a good level of intelligence. I was not disruptive but easily bored by lessons. I preferred creative activities because there were usually no limits to where that activity could go. My levels of attainment were average despite having a good brain; as a student I had lacked focus and struggled to apply myself. My chemistry teacher at the time wrote me off as lazy. (I didn't hold him in high enough regard to explain I regularly worked 30 hours a week, supplementary to my studies.) My Biology and Physics teachers were inspiring (thank you Ms Forrest, Mr and Mrs Kleinhans).

A Different Choice

When the time came to apply for university I was clutching at straws as to what direction I was to take. My exam performance was poor, and issues at home had taken a heavy toll on what had started out as predicted exam results on the upper side of average. I knew that after college I wanted to progress in life but needed to earn money. And, as much as it is hard to admit, I didn't have the self-confidence to go away to a University. More importantly I didn't have the grades to gain entry to a University degree or with a remotely competitive admissions profile. I knew my future plans for University would begin with the 'clearing' process. I may have lacked focus and application but I would be damned if I'd let the starting course of my future as an adult be decided on what empty gaps there were in a University intake. I sought the opportunity to discuss what I might try and do next with a career advisor with the local education authority. Her name was Gill Pimblett and she didn't give me any advice. Her approach was much better than that — she gave me options to choose from. I wanted to stick with what I knew best so decided to try and secure an engineering/science apprenticeship. Nowadays, apprenticeships have had a resurgence. Then (in the late 1990s) my fellow students had no idea they still existed. I made applications, did some aptitude tests, and went for my very first selection interview in a grubby training room at a fertilizer plant. It did not go well. The day after, I got a phonecall offering me an interview in Runcorn for a Glassblowing job.

I decided why not. It was on a busroute and it sounded cool. I was still recovering from my first interview at the fertiliser plant. I'd gone into that with half a plan of how to act and present myself. It hadn't worked; I decided to go into the glassblowing interview with no plan. Probably my first 'best' decision. Arriving at ICI's Heath headquarters, I went into the coldwar era laboratory blocks and presented at reception. I was given my very first visitors induction, security pass and access card by an intimidating woman and told to wait in the reception area for someone to come and get me. I was out of my depth.

The Interview

I sat a round table with my back to the door, facing two men in spectacles. They were both short, and in their late 50s. One had a beard and he was stroking it. The other, a moustache. They were both eyeballing me. A pencil and paper lay on the table, along with a piece of glass apparatus. It was I would later learn to be a Davies double surface condenser. The one with the moustache asked me to draw it. I was observant enough to tell there were parts inside, so I drew those as well. That turned out to be a good idea. I finished the drawing. They then asked me what my specialist subjects were. At the time I had no idea what they were talking about. The one with the beard was a historian and a musician. The one with the moustache was a draughtsman/illustrator and musician. They took me to the glassblowing workshops and I had a go, I melted some glass and didn't burn or cut myself. The day did not feel like a success. I liked the idea of glassblowing and I left desperately wanted to work with those two men.

I got the job.

The Tea Club

At the height of its power, the British empire was the largest empire in human history. At its most extensive, its domain encompassed a staggering 25% of the earth's land mass, and within it resided around 458 million souls. The legal, cultural, linguistic and geographical diversity of the British empire required a feat of administration, and all this in the days before the fibre optic cable. It was, however, nothing compared to Howard's tea club.

Howard Bancroft practised a discipline that I call Hostile Administration. Each of us were required to pay £1 per week into the tea club. For this, you were allocated a cup of tea of coffee at each break time. Your contribution entitled you to single teaspoon of sugar should you require it. Sweeter tooths were not provided for. A row of re-dried tea bags dangled from pegs in the workshop — they were used as a reminder of how the teaclub dealt with dissidents.

The Tea Clubs fiscal policy did not make provision for biscuits. However, a host of grateful customers kept us well supplied and their rationing was controlled by the tea club. The allocation was one per break time.

This was a robust system, administered using a single A5 margin ruled notebook (black hardbacked with a red spine) and an enamelled metal tin. Howard used a set square to rule columns into the book for logging tea club member's contributions. Contribution day was a Wednesday.

I would always contribute but would often forget to bring a £1 coin. This was the preferred payment — smaller denominations of coinage would be taken upto a point, with a degree of revulsion. Notes even more so. Regardless of the reason, non contributors were marked in the book with "UNPAID" against their name. In RED pen.

Once, I was entrusted to put a bag of sugar for the club. Somehow I manage to buy a bag that had at some point gotten wet — and I presented Howard with a single 1kg lump of sugar. I was made to take it back and exchange it, and given specific instructions to get an exchange rather than secretly buy another bag with my own money. At the time I was nervous about going back for a refund (I was not a confident person). I did it though. This was my first lesson in vendor management & being an intelligent customer.

Howard's dedication to due process was the kind of behaviour that made controlling the British Empire possible. At its peak, the 20% of the world's population were within the empire. Our tea club had three members.

The term 'Hostile Administration' appeals to my sense of humour, but in truth this was a valuable lesson in systematic admin and consistency in approach.

I learned that Howard was a creature of habit — and these habits were one of the tools he used to maintain high standards. His Tea Club system was a relic from bygone days of managing a workshop with 8 or 9 members of staff — but it was scalable and it was fair.

He employed the system without prejudice or favour, and it worked. When I became a manager of a bigger team, and gained insight in the tools and methods of corporate HR policy I learned that systematic working and consistency in approach are what make human systems work.

Craftsmanship, Service and Hygiene

One could write up some corporate style vision statement but for a glassblower, what this basically amounts to is: "We need to consistently turn out good, solid glassware the way the customer wants it, on time, and we need to make sure we don't turn the workshop into a complete dump while we're going about it."

Keep a Notebook

In my first week I was told to buy a notebook and write up each week what I'd been working on and learning. For me it was a composite of work instructions, drawings and a diary. I still have it. This is useful for two reasons: firstly, to look back at what progress has been made over the weeks and months, and secondly to revisit on technical aspects/finer details. I still keep a notebook.

Be Nice to the Canteen Lady

Best lesson I ever learned, as I recall this was told to me on day one. Be nice to the canteen lady. The canteen lady was of paramount importance (she was in charge of food). This amounted to a lesson in courtesy and respect, and how to identify key people in the organisation. I took it to mean that respect isn't something that should be reserved for senior people in the organisation. I think it also meant that everyone has their own sphere of influence and you'd be wise to learn how to recognise it.

Appraisal, Critique and Evaluation

The master and apprentice relationship should involve regular assessment of the work produced, from continual direct supervision and critique at the start, progressing to evaluation of the workpieces; the frequency of evaluation should typically diminish as apprentice's skills and the complexity of the workpieces develop. I believe it is paramount that any assessment is completed with feedback given. Feedback must be honest, open, supportive and varied — it should take in all aspects of the work, good and otherwise, and be given in earnest. That's the way it worked during my apprenticeship.

Let's not forget glassblowing is hard to learn, so whilst mollycoddling must be avoided like the plague, one should bear in mind the emotions and sensitivities of an apprentice and use a style that will get the best out of them.

For this to work for any apprenticeship though, the apprentice has to accept the contextual superiority of the master; an apprentice needs to be confident in the master's abilities, and aspire to them, otherwise there is potential for incredulity or even resentment to affect the relationship at some stage. I was lucky enough to work with a Master whom I still aspire to emulate. Of course, the level of potential for negative effects is rooted in the personalities of the master and the apprentice. For that reason, the concepts of Critique, Evaluation and Appraisal should be borne in mind at the recruitment stage. Individuals who can't respond to feedback won't last long. They'll either burn or cut themselves, or just not turn in one day.

Critique is an assessment of the all aspects of the work produced and the approach taken. The idea is that this is a part of the learning process and should be fairly frequent — personally I think as frequent as the lessons given. Critique is a compass for an apprentice to find their way. It needs to be tuned to the standards that are required at that stage in the learning process.

Evaluation is an assessment of the work produced or the methods used against the standard required. These standards can either be method/behavioural standards or technical standards. Evaluation involves more structure than critique. An example might be a skills test where a piece of apparatus is produced against a drawing. The master will evaluate the work produced by taking measurements and comparing with the drawing i.e. assessment against the standard required.

To me, Appraisal should be a composite of Critique and Evaluation. It's an assessment of the value of the apprentice's skills and behaviours based on opinions and observations (as in critique) validated against quantified or demonstrated standards (as in evaluation).

Here's a 10-second example: "Your work is exceptional — its exactly as it should be in against the drawing. But your work bench is a mess and you've don't put your tools away." Standards and behaviours.

The conversations I have with my apprentice master still on occasions involve a degree of evaluation. Since the start of my apprenticeship we've used a simple scoring system of marks out of ten. To this day, I have yet to be awarded a score greater than 3/10. 3/10 is fine — on our grading scale that's a baseline standard of competence. We both know what success looks like. The score was far less important to me than the level of interest he took in my work. The time he took a select few of the dozens of U-tubes I'd be practising making and packaging them for delivery to a customer was one of the first times I felt like a success.

Imitation, Style and Improvisation

This worked for me because a craft apprenticeship relies heavily on learning through imitation; imitation of the master's technique and approach is the mainstay of developing basic ability in manipulating a material. At the point where the apprentice begins to improvise outside of the imitated actions and behaviours is where their signature style begins to develop.

Molten glass is probably the most difficult material one can work with. It comprises the technicalities of hot metal working, with the artistic challenge of sculpture, while one fights gravity. It is this composite nature of the medium that makes this regular evaluation of the work produced.

Relationships, Philosophies and Behaviours

The Apprentice Relationship

I've spoken to a few people over the years who've been lucky enough to have a real apprenticeship. Though similarities can be drawn, a relationship between an apprentice and master is unique and like no other.

A good apprentice–master relationship is much closer than teacher-pupil in that attention and support from a teacher is shared between a class of pupils in varying measures. Most of the instruction is task focussed. Success is generally judged rather than measured; observation on the basis of exhibited standards, competencies, and behaviours progress and success can be determined. You could summarise that into a word: Trust.

And trust, of course, must be earned.

Trusted to work on your own. Trusted to talk to a customer. Trusted to cover the workshop while the master is in a meeting or with a customer. Trusted to complete an order. Trusted to make a decision.

To an apprentice, success is being trusted.

Masters and Mentors

It would appear that there are fewer and fewer opportunities to use the term 'Master' correctly, and perhaps the term 'mentor' suits a more contemporary approach. Personally I see value in the definition of "Master and Apprentice". A mentoring relationship involves a person of experience (the mentor) and a person seeking experience/learning (a protege if you're that way inclined, or apprentice, or 'mentee' in corporate speak). A Master on the other hand, is not just someone of experience; a Master is, or should be, someone who has truly mastered their craft. This is a fundamental difference.

A Mentor can give advice, guidance and the benefit of experience. A good mentor can identify and develop potential in their protege that they the mentor do not themselves possess.

A Master can actually demonstrate what competence looks like; a competent Master consistently delivers high standards through multiple methods far beyond those required for a deemed competence, within the limits of the material or subject.

To truly develop and grow talent in an apprentice (especially a glassblower) you need to be both a Mentor and Master.

My perspective is that to be a mentor or Master, success is seeing evidence that your mentee has listened to your advice and considered what you've said. Success is observing effort and commitment in a response to one's own efforts. Success is seeing a mentee take due pride in achieving a standard. Success is seeing a mentee develop and exhibit new behaviours and attitudes.

You could summarise that into one word: Respect.

And respect, of course, must be earned.

Respect for the art / craft. Respect for the task or the objective. Respect for the work, the tools and the work environment. Respect for the customer. Respect for the relationship. Respect for the Master.

I do not know if by my own definition I am a Master nowadays in the field of expertise I have settled in. Vanity makes me like to think I am. But I am fortunate enough to be a Mentor. I am even more fortunate to have been chosen as such by a 'mentee' who I am working with on my current employer's First line management programme. Reflecting on my experience as an apprentice has shaped the way I'm working with my mentee. We're not dissimilar in age separated by 5 years, but we've vastly different backgrounds in terms of career history and experience.

It's not easy. Frankly, I haven't got the free time to Mentor someone, but I make time as much as possible. I always appraise in earnest anything I'm asked to give an opinion on. I also put a lot of effort into observing the standards my mentee is demonstrating. Having said that, I am also a fairly demanding person with standards that I won't compromise on.

The reward for me is observing the commitment my mentee has shown — in listening and acting on the advice and opinions I've given, in continually working hard and delivering on projects and activities that I have handed over, knowing fullwell that there would be elements of difficulty and that the attributes needed would not come naturally to her. Mentoring this person who has committed themselves to improving and developing has been one of the most enriching experiences of my now matured career. And not the least it's been a chance to pay something back.

Philosophies and Behaviours

Mentoring is a privilege. It is an opportunity to use one's experience and philosophy to try and make a positive influence on an individual seeking to develop themselves. For that reason it's important to enter a mentoring relationship with at least an understanding of the philosophy you employ and a degree of consistency to your behaviours and standards. Mentoring must be taken seriously, and before one attempts to shape the philosophies and behaviours of an apprentice they must reflect on their own and be sure they are sound. I think learning how to do this was probably the most valuable lesson I learned from my apprenticeship.

If you can't do this then you are not upto the job. The good news is, this need not be complicated

Behaviours: Practice Makes Average

This phrase was discussed a lot in the workshop during my apprenticeship. The lesson could be either reassuring or soul destroying depending on one's level of aspiration, but it is nevertheless true. Basically, regardless of the skills gap, if you practice something for long enough, you will at least attain an average level of skill, i.e. a basic level of competency. In the case of glassblowing this should be reassuring as when a complete beginner sets out on developing their skills, weeks can pass by with a feeling of getting nowhere. There was a second lesson to this though; practice will only develop so much talent, refinement of skills takes more than just patience and repetition. Refinement of a basic or even a good competency into an expertise requires a level of engagement with the craft and discipline far beyond repetitive learning. For most glassblowers the circumstances of the work passing through the workshop means that a step up in this engagement is natural; Customers want and need to talk to the person making their apparatus. As trust grows and an apprentice's remit increases the level of engagement steps up from repetitive bench working to customer engagement, eventually to full project management. It is at that point when the apprentice (or even a glassblower deemed as competent in producing work to a drawing) begins to interact with customers that behavioural factors can decide the fate of a glassblower, which leads us to the next lesson.

Behaviours: Ask the Right Questions

Modern business uses the strapline "Right First Time". This means let's not waste money or time or annoy our customers by making a mess of things.
If we ask the right questions when a job is being briefed, we can save everyone a lot of hassle. This is one of the behaviours that makes the difference between a factory worker and a service provider. A good glassblower is a provider of expert service. Here's an example:

An organic chemist walks into your workshop with a neat drawing that says the bubbler apparatus they want is 248 mm long, 50.2 mm wide and the connections on the side of the bubbler are defined as 8.35 mm in diameter.

Experience tells us the bubbler is basically 250 mm long, it's made from 50 mm tubing but what about the connections? 8.35 mm is pretty specific. It MUST be important.

So what questions do we need to ask.
Do you have a spec, for the precision ground 8.35mm tubing?
Can you give us your required lead time?
Do they realise the cost implication?

No. the most important question is: What do you do with the connections on the bubbler? Answer 99/100 times: "I shove rubber tubing on it from my gas supply."OK — standard 8 mm tubing it is then.

Conversely. A Physicist walks in to your workshop with a laboratory caricature that looks like they've drawn it on a McDonald's bag, with their feet, using a crayon that has at some point been snapped and repaired with sellotape. Around the crayon there is a lot of scribbly writing that has been most certainly written using a fisher space-pen, there are lots of numbers to three decimal places (d.p.) and something is labelled as 'very flat'. Chances are the need for three d.p. is legitimate and the "very flat" bit is probably an optical window. That's when you ask a lot of questions about specification and go to their lab to see what the application is with your own eyes.

Maintaining Standards

The Paper Necktie Perception is Everything

After being recruited into Glassblowing I was sent on a laboratory apprenticeship scheme for a year. I came back to the workshop for a visit one day, wearing a yellow t-shirt and jeans. A photocopy of a necktie was promptly sellotaped to my chest and there it was to remain for the rest of the day. I did not turn up again without a collar and tie. We're glassblowers not fitters, I was told. We wear a collar and tie with a Windsor knot. It was explained to me what a Windsor knot was. Looking back this was about perception and standards — and how to differentiate one's self. A valuable lesson that I complied with, but didn't fully appreciate at the time. In my more recent career the importance of managing perceptions has moved a lot higher up my list of priorities.

This lesson was firmly rooted in the machinations of a corporate industrial entity (which in this case was ICI plc.) within which it is highly unlikely that any glassblower's will ever train or work again, the industrial climate being what it is nowadays. Those glassblowers working in the creative industries and Academia will probably have no frame of reference for this, and it is likely to be a lesson that isn't relevant to the contemporary glassblowing industry. I've included it in the hope that the glassblowing industry will rejuvenate to an extent that it again becomes relevant in some form. It's one that I forgot over time, and am glad to recall now. At this stage in my career it's more important than ever.

On a practical note though, I was horrified and still am that I wore a necktie while using unguarded rotating machinery. I never forgot to tuck the tie into my shirt when working and had no incidents occur but I still shudder to think about it.

Appropriate Specification

I noticed early on that the glassware we produced was heavier and more robust than catalogue items from the main manufacturers. This was down to the medium-wall tubing we used (a heavier thickness of material) as opposed to the more commonly used and thinner/cheaper light-wall tubing. Borne out of the ICI Alkalis/Caustics division that my apprentice master had come from, we always build equipment to last; caustics essentially eat glassware from the inside out, and thicker tubing meant apparatus lasted longer. I was pleased with this approach. It meant we produced robust apparatus that had a quality feel, but didn't look clunky or over specified. It was cost efficient, we retained customers well because our apparatus had a reputation for being durable, and the workshop wasn't inundated with repetitive repair work on needless breakages.

Standards of Communication

Our workshop had the privilege of working for British Nuclear Fuels on some interesting projects. I was entrusted with keeping the workshop ticking over when my colleagues were on leave. I took a call from BNFL who described a modification they wanted to some apparatus we'd routinely make for them. I asked for the drawing number and talked the spec through — the distance between centres, sizes etc. I specifically remember my poor choice of words on the day, which were "I understand. You want me to put a bit of a kink in the outlet tube. Not a problem." I was met with a few seconds silence, then came a response "No. We require a controlled bend." Understood.

This goes back to perceptions again, and choosing standards and styles of communication to suit. I quickly learned how to turn formal communication styles off and on quickly to suit my needs. Conversely, I got very good as swearing (I hardly ever swore before) as a way of fitting in with men in the engineering workshops — something I was very apprehensive about when I was starting out. Horses for courses. The key to doing this properly is not mixing your styles up and using them at the right time in the right situation on the right people.

Housekeeping and Hygiene

If you've ever made a hole in a price of glass by spot heating and blowing out you'll understand how hilarious it is to see a bubble of glass only microns thick appear out of nowhere. Then comes the fun of bursting it or knocking it off with a carbide cutter. This novelty didn't wear off for me. EVER. I combined this juvenile enthusiasm with a very conservative attitude towards waste. As such my workbench was a complete tip and it encumbered me on a daily basis. Housekeeping standards are as important as craftsmanship. To illustrate this point I was sent to work in a basement workshop at another site, with a very nice and skilled glassblower who shall remain nameless. I was there for two weeks and I learned that he too did not like to throw things away. Nor did he like to brush up. Imagine being inside one of those carnival stalls where people pay to throw bricks at glass and crockery. In the semi dark. With fires. When I came back my standards improved lesson learned.

Never Stop Learning, and Never Stop Sharing

I remember being told about the secretive nature of bygone glassblowers, who coveted knowledge and skills and greedily hid them so as not to weaken their position, as if in some way they could ensure their ongoing value if they made glassblowing appear to be some kind of Druidic black art. Bit pointless really as glassblowing is so hard to master. Nevertheless I understand that in some circles this behaviour was commonplace. Luckily I've never seen it for myself. I've been fortunate enough to study with glassblowers in the USA, Europe, and New Zealand, and everyone of them shared their knowledge freely, with enthusiasm. Perhaps even more encouraging for me was their preparedness to entertain new ideas and techniques that I had to share, even with my modest skill set. That generosity is something that I've never forgotten, and even though I'm no longer in the field I still try to pay that back in as many ways as I can.

When it comes to learning, if there's one thing you'll take from this text, take this: Be generous and listen in earnest.

Index

A

A Contemporary Art, 383
A double-walled flange, 400
anneal the glass, 378
annealing oven, 402
annealing point, 1
annealing, 3
annular space, 167, 349
appraisal, 423
artistic approach, 383
artistic glassworking techniques, 379
Asbestos Paper, 200
Ask the Right Questions, 428
atmosphere, 337
atmospheric pressure, 335

B

90° bends, 81
bench burner, 130
bends, 191
bends, 90° and U, 130
blow out, 361
borosilicate flat plate, 295
borosilicate glass, 25, 27, 210, 408
borosilicate to fused silica, 268
borosilicate to soda lime glass, 268
bourdon gauge, 338
brace the seal, 132
Brasher method, 348
brass reamer, 101
bubble plate column, 355
Buchner flask, 336
bulb blowing, 149
bulbs, 191
burners, 12
button, 168
button seal, 100, 134
buttress joints, 220
buttresses, 244

C

CAD, 374
calibrations, 299
capillaries, 408, 417
capillary joins, 186
capillary tubing, 84
carbon dioxide, 19
carbon paddle, 380, 388

carbon reamer, 105
carbon/graphite points, 288
carborundum, 233
carborundum lapping machine, 224
carborundum powders, 248
challenging goals, 400
chucks, 126, 129
coefficient of expansion, 254
coefficient of thermal expansion, 2
coil condenser, 401
coil winding, 206
cold working, 383
cones and sockets, 212
control the blow, 64
controlled blow, 129, 160, 196
copper to glass, 265
copper wire, 201
Corian®, 394
cork borer, 59
cork to the ceiling, 88
corrugated carbon, 331
corrugated card, 110
corrugated graphite tape, 201
corrugated paper, 200
critique and evaluation, 423
cubic centimetre, 299
cup sockets, 244
cut off, 391
cut-off machines, 221
cutting flat glass, 44
cutting of glass, 33

D

Davies double surface condenser, 271
definition of a standard, 367
deionised water, 348–349, 374
devitrification, 371
Dewar seals, 216, 236
Dewar vessel, 241

diamond impregnated discs, 248
diamond lapping machines, 223
didymium glasses, 193
didymium safety glasses, 13
distillation-head, 399
distilled water, 348–349
dome, 74, 79
domed end, 57
domed or test tube end, 73, 89
dominant hand, 63
double re-entrant, 283
draw down, 113, 118
Dreschel Head, 109
dual manifold Schleck line, 322
dwell time, 4

E

effects of gravity, 49
emergency lighting, 20
evaluation, 424
expansion bellows, 174
expansion/contraction, 409
extended sockets, 216
external, scroll ring, 126
extraction systems, 18
feather-like flame, 78, 171, 199, 207–208, 332, 360, 362
fibre glass, 200
fire alarm sensors, 20
fire alarms, 20
flame anneal, 90, 199, 412
flame control, 77
flame polishing, 33, 38, 49, 109, 173, 175, 374
flammable fuels, 18
flange making, 236
flat carbon paddle, 129
flat carbon plate, 138
flat flanges, 220, 244

flat lapping machines, 223
flat spiral, 164
float glass, 224
flooding rate, 354
Foster cell, 389
fractional distillation columns, 353
fractionating column, 400
freeing of joints, 292
freeing oil, 292
fuel gasses, 18
fused silica, 27

G

gases, 12
glass mini-mixer settler, 384
glass retorts, 353
glass to metal seals, 254, 408
glass working tools, 14
glassblowing lathe, 14, 124
graded seals, 268, 408
graphite rod, 101
greased high vacuum taps, 339
grip of the wheeled cutter, 45

H

hand torch, 129, 312
heat-resistant woven cloth, 201
hot blow out, 313
hot point, 113, 331
housekeeping, 21
hydrofluoric acid, 374

I

imitation, 425
indentations, 287
industrial diamond, 221
initial join, 63, 71, 73, 77, 189, 196

initial ridge, 364
International Organization for Standardization (ISO), 367, 370
interwarming, 327
ionisation gauges, 338

J

jigs, 384, 395, 397

L

laminated glass, 48
laser cutting, 374
lathes, 225
Lavoisier, Antoine, 211
left hand dominant, 65, 152
Liebig condenser, 167, 353
Liebig, von Justus, 167
lifting equipment, 20
lighting, 20
Linisher belt grinding, 225
Linisher machine, 229

M

maintaining standards, 429
mandrel, 163
manometer, 95
masterpiece, 398
McLeod gauge, 283, 338
mechanical rotary pump, 338
melt back hole, 361
metal mandrel, 161
methyl salicylate, 292
millibar, 337
millilitres, 299
mixer settlers, 384
molecular beam system, 408
molecular pumps, 339
Monel random packing, 354

N

never stop sharing, 431
nickel–iron–cobalt alloys, 264
nitric acid, 378, 417
nitrogen dioxide, 19
nitrous oxide, 19
neon pulse, 20

O

Olive exercise, 136
olives, 135, 140
one melt side arm, 139
optical window, 411, 408
orange sodium flame, 63
over blow, 65
overblown, 95
over hand grip, 63
oxygen, 341
oxygen cylinders, 17
oxygen enrichment, 18
oxygen/gas flame, 313

P

packed fractionating column, 353
parallel coil, 166
Pascal, 337
Pasteur, Louis, 121–122
pause, 77
pear-shaped column, 156
pen grip, 76
personal protection equipment (PPE), 21
philosophies and behaviours, 427
photoreactor, 371
Pirani gauges, 338
pitch, 163
platinum, 255
plunger type taps, 340

polarised light, 49, 322
poly-tetra-fluoro-ethane (PTFE), 24, 385
practice makes average, 427
preheating glass, 383
prewarming glass, 327
prewarming, 331
Priestley, Joseph, 211

Q

quartz, 371

R

rectangular cross-section, 387
refrigerated liquid oxygen, 17
ribbon burner, 93, 130
right angle bend, 77, 88
right hand in support, 65
ring seal, 391
rounded shoulder, 66, 71
run in, 77, 80, 89, 103, 129, 171, 173, 178, 244, 256, 362

S

safe working pressure, 65, 250
Schlenk lines, 339, 377–378
scientific glassblowers, 379, 383
scratch marks, 27
sealing holes, 292
sensitising solution, 349
shoulder, 68, 73
side arm, 89, 105, 178, 183–184
side seal, 175, 183
side-tubes, 392
silver chloride, 348
silver nitrate, 348
silvering solutions, 348, 350
sinters (filters), 236

skirted cone, 216
soda glass, 27, 54
soda-lime glass, 210, 255
sodium flame, 130, 168, 176, 314, 326, 362
soft flame, 313
softening point, 1
solvent extraction, 385
Soxhlet top re-entrant, 284
spear point, 49, 59, 68, 100, 103, 121, 151
specific gravity bottle, 229
spherical joint, 219, 416
spiral condenser, 175
spiral winding, 161
splash head, 151
spring loaded chucks, 126
stainless steel, 266
standard kilogram, 299
standard of competence, 425
standards, 424
standards of communication, 430
star crack, 27
straight join, 59, 63, 76, 124, 360, 362
strain, 198
strain patterns, 49
strain point, 2
stress relieve, 198
stresses, 383
strong in compression, 28
style and improvisation, 425
suck seal, 279
suction cups, 47
sun and planet, 126
support tube, 391
supported internal seal, 110, 167, 200
surface tension, 139
swivel or bend, 312

swivel, 187, 191
synthetic mullite, 267

T

T piece join, 171, 326
T pieces, 74, 133, 191, 196
tailstock, 391
tapered shoulder, 73
tesla coil, 345, 408
test tube, 57
tetrafluoroethylene hexafluoropropylene copolymer, 220
theoretical plate, 353, 355
thermal shock, 27
thin bubble of glass, 74
thin bubble, 106
three neck flasks, 357
timing is everything, 83
Torr, 337
Touch–Pull–Blow, 63, 106
toughened glass, 48
tungsten carbide glass cutter, 33
tungsten pin, 411
tungsten sealing glass, 416
tungsten to borosilicate, 258
two-point seal, 321

U

U bend, 91, 95, 180, 318
U bend manometer, 98
ultraviolet (UV), 371
unequal joins, 66
UV light, 389

V

V block, 42, 65, 109, 134–135, 141, 161, 230
vacuum, 250, 335

vacuum 10–10 torr, 419
vacuum measurement, 336
vapour flow rate, 354
Veigreux column, 287
Vernier or digital calipers, 229
volume of a cone, 296
volume of a cylinder, 295
volume of a sphere, 296
volumetric ware, 295

W

wall thickness, 65
warming process, 378
water jet ejector, 338
water jet pump, 134
weak in tension, 28
wired glass, 48
withdraw and pause, 65
wooden supports, 394
workbench, 12
working temperature, 1
woven fibre glass, 200

Y

Y pieces, 77, 132

Lightning Source UK Ltd.
Milton Keynes UK
UKHW020603250822
407819UK00001B/49

9 781786 341976